COMPUTER GAME DEVELOPMENT AND ANIMATION

PRACTICAL CAREER GUIDES

Series Editor: Kezia Endsley

Computer Game Development & Animation, by Tracy Brown Hamilton

Craft Artists, by Marcia Santore

Culinary Arts, by Tracy Brown Hamilton

Dental Assistants and Hygienists, by Kezia Endsley

Education Professionals, by Kezia Endsley

Fine Artists, by Marcia Santore

First Responders, by Kezia Endsley

Health and Fitness Professionals, by Kezia Endsley

Information Technology (IT) Professionals, by Erik Dafforn

Medical Office Professionals, by Marcia Santore

Skilled Trade Professionals, by Corbin Collins

COMPUTER GAME DEVELOPMENT AND ANIMATION

A Practical Career Guide

TRACY BROWN HAMILTON

ROWMAN & LITTLEFIELD
Lanham • Boulder • New York • London

Published by Rowman & Littlefield
An imprint of The Rowman & Littlefield Publishing Group, Inc.
4501 Forbes Boulevard, Suite 200, Lanham, Maryland 20706
www.rowman.com

6 Tinworth Street, London, SE11 5AL, United Kingdom

British Library Cataloguing in Publication Information Available

Library of Congress Cataloging-in-Publication Data

Names: Hamilton, Tracy Brown, author.
Title: Computer game development and animation : a practical career guide /
    Tracy Brown Hamilton.
Description: Lanham : Rowman & Littlefield Publishing Group, 2019. |
    Series: Practical career guides | Includes bibliographical references. |
    Summary: "Computer Game Development & Animation which includes
    interviews with professionals in the field, covers the following areas
    of this field that have proven to be stable, lucrative, and growing
    professions. Artist/Animator. Producer. Sound Designer. Video Game
    Designer. Video Game Developer. Video Game Tester. Writer"— Provided by
    publisher.
Identifiers: LCCN 2019038846 (print) | LCCN 2019038847 (ebook) | ISBN
    9781538133682 (paperback) | ISBN 9781538133699 (ebook)
Subjects: LCSH: Computer games industry—Vocational guidance.
Classification: LCC QA76.76.C672 H35165 2019 (print) | LCC QA76.76.C672
    (ebook) | DDC 794.8023—dc23
LC record available at https://lccn.loc.gov/2019038846
LC ebook record available at https://lccn.loc.gov/2019038847

♾™ The paper used in this publication meets the minimum requirements of American
National Standard for Information Sciences—Permanence of Paper for Printed Library
Materials, ANSI/NISO Z39.48-1992.

Contents

Introduction   vii

1   Why Choose a Career in Computer Game
Development and Animation?   1
2   Forming a Career Plan   19
3   Pursuing the Education Path   37
4   Writing Your Résumé and Interviewing   57

Notes   71
Glossary   75
Resources   79
Bibliography   83
About the Author   87

Introduction

So You Want a Career in Computer Game Development and Animation

*W*elcome to the field of computer game development and animation! This book is the ideal start for understanding the various careers available to you within the computer gaming industry, which is right for you, and what path you should follow to ensure you have all the training, education, and experience needed to succeed in your future career goals.

Computer game development is an exciting, creative, and thriving industry, and also a broad one. There are many roles and careers that fall under the umbrella of computer gaming—from developers to writers, animators to testers, and everything in-between. Having so many paths to choose from is exciting, but it can also make it difficult to choose which is the most fitting to you.

A Career in the Computer Gaming Industry

Covering every career path involved in the creation of computer games is beyond the scope of this book. It will however cover the most common. This book will cover careers including

- Artist/animator
- Producer
- Sound designer
- Video game designer
- Video game developer
- Video game tester
- Writer

These jobs are widely available pretty much all over the country. They pay pretty well, too, considering that they don't all necessarily require that you have a college degree. And the industry is booming. In 2018, the gaming industry had massive growth, thanks to many new products and advances in technology. It was predicted the industry would be worth more than $130 billion by the start of 2019.[1] That's a very healthy outlook and good news for anyone looking to enter one of these professions. The future looks bright for these jobs as well, as you'll see.

The Market Today

How does the job market look for young people seeking to enter the field of computer gaming? Although the field is a competitive one, it is growing rapidly. Advances in technology—including not only software advances but also the prevalence of devices such as smartphones and tablets on which people can play games more easily on the go—have kept the popularity of computer games growing, and the sophistication of the games is increasing continuously.

Employment opportunities in the field of computer gaming are expected to increase substantially over the ten years between 2016 and 2026. Video game designer opportunities are predicted to grow by 30 percent during that time period according to IT Career Finder.[2] That's certainly a positive prediction—and a greater increase than most professions, which is around 7 percent.

Already a booming industry—it has tripled over the last decade—careers for game designers, artists, writers, and producers, and so on are expected to grow substantially. Mobile game development—creating games to be played on smart devices—and more sophisticated technology development has created a demand for games incorporating augmented reality (AR) and virtual reality (VR) technologies.

In recent years, the number of gamers has steadily grown. In 2018, 66 percent of the US population over age twelve were gamers. That's up from 58 percent in 2013, according to Nielsen.[3] The market is predicted to be worth a staggering $90 billion by 2020—an increase from nearly $79 billion just three years earlier.[4]

A career in the computer gaming industry offers a broad range of creative and diverse roles. *iStock*

Several factors account for this industry growth, which far exceeds the growth percentage of other careers. According to the *Motley Fool*, these factors include

- Greater creativity, adding twists to traditional games
- More inclusiveness with the development of games that appeal to a broader, female audience
- Better graphics, with improvements that make gameplay more realistic and satisfying
- Game streaming, enabling people to observe and learn about new games easily
- Nostalgia, with classic games returning to popularity
- Massive growth in the number of gamers worldwide[5]

Many of the top employers in the field of computer gaming are located in the southern states—with seven of the top ten locations with the best opportunities in the field being in the South, where salaries are high and the cost of living low—as well as technology hubs like Seattle and San Francisco. According to some research, the following cities in the United States will offer the best opportunities for high-paying, secure jobs in the computer game development field:

- Austin
- Atlanta
- Raleigh
- Columbus
- Seattle/Bellevue
- Lexington
- San Francisco
- Durham
- Dallas
- Huntsville
- St. Louis
- San Jose
- Phoenix
- Cincinnati
- Santa Clara
- Sunnyvale
- Lawrence
- Orem
- Chapel Hill
- Boulder[6]

But if you don't live near these cities or don't see yourself wanting to in the future, don't despair: Freelance, contract, and remote opportunities also exist, from the smallest towns to the biggest cities, and all over the world.

From Streets of Rage to Halo to World of Warcraft, I was always enthralled by these worlds, their communities, and the stories within them. Additionally, the added challenge of audio implementation gives so many opportunities for creating incredible experiences.—Elliot Callighan, composer and sound designer for games

What Does This Book Cover?

This book covers the following topics for all the aforementioned careers, as well as others:

- What kind of job best suits your personality and preference for working conditions, hours, educational requirements, work culture, and atmosphere, including the day-to-day activities involved in each job and what a typical day at work will look like
- How to form a career plan—starting now, wherever you are in your education—and how to start taking the steps that will best lead to success

- Educational requirements and opportunities and how to fulfill them
- Writing your résumé, interviewing, networking, and applying for jobs
- Resources for further information

Once you've read the book, you will be well on your way to understanding what kind of career you want, what you can expect from it, and how to go about planning and beginning your path.

Where Do You Start?

All the jobs we cover in this book require at a minimum a high school degree or equivalent, such as a general education development (GED) certificate, and some on-the-job training. Others require a four-year degree and even a master's degree—and the subjects you can follow will also vary. Computer science, art, programming, and creative writing degrees are all relevant to a career in computer gaming. And if you want to start your own company or run an existing one, business degrees are also recommended.

A lot of choosing the right career for you will also depend on your personality and interests outside of work—such as whether you work better with people or independently; whether you want to be the boss or work for someone you admire; what you want your life to include outside of working hours—including hobbies and other activities that are important to you; and so on.

After high school, knowing how to choose and apply to vocational training such as an apprenticeship or a college program will be the next step in your path. The information in chapter 3 will help you navigate this important stage and know what questions to ask, how to best submit yourself as a candidate, and the right kinds of communication skills that are key to letting future employers or trainers understand who you are and what your potential is.

Thinking about the future and your profession is exciting and also a bit daunting. With this book, you will be on track for understanding and following the steps to get yourself on the way to a happy and successful future in the computer gaming industry. Let's get started!

Why Choose a Career in Computer Game Development and Animation?

The fact that you picked this book off the shelves and are reading it means you have decided you are interested in taking your passion for creative writing, animation, programming, or simply gaming to the next level: considering computer game development and animation as a career. Choosing a career is a difficult task, but as we discuss in more detail in chapter 2, there are many methods and means of support to help you refine your career goal and hone in on a profession that will be satisfying and will fit you the best. Of course the first step is understanding what a particular field—in this case computer game development and animation—actually encompasses and informing yourself of how the future outlook of the profession looks. That is the emphasis of this chapter, which looks at defining the field in general and then in more specific terms, as well as examining the past and predicted future of the field.

Computer gaming development, as mentioned in the Introduction, involves any number of different but related roles, all of which are necessary to create fun, challenging, impressive, and popular computer games that will be played by avid gaming enthusiasts around the globe. However, each career this book covers—from the person writing the game's narrative to the person creating the graphics to the person testing the beta version along the way—requires creativity, flexibility, attention to detail, various levels of technical skills, and a passion for gaming. And because so many people working in the various fields may have slightly different educational backgrounds or ideas, the field requires excellent communication skills, the ability to collaborate, often the willingness to work outside of a nine-to-five mind-set, and flexibility.

The desire to be constantly learning is also a benefit, as any job relating to technology—a fast-moving, frequently changing world—demands a curiosity to learn new technologies and apply them in the best way. If you are truly passionate about gaming, a career in computer game development and animation will be continuously satisfying. It is also a very competitive field, often demanding long hours and entailing working in high-pressure, often stressful environments. Although the field is growing quite impressively, there are many talented, creative, and motivated people vying for positions in what is one of the most creative and fun jobs out there. Game development is also a field that demands the ability to face tight, crunch-time deadlines.

So as with any career, there are pros and cons, which we will discuss in the chapter. In balancing the good points and less attractive points of a career, you must ask yourself whether, in the end, the positive outweighs any negatives you may discover. This chapter will also help you decide whether a career in computer gaming development and animation is actually the right choice for you. And if you decide it is, the next chapter will further offer suggestions for how to prepare your career path, including questions to ask yourself and resources to help you determine more specifically what kind of career related to computer game development suits you the very best.

What Is Computer Game Development and Animation?

As mentioned previously, the computer gaming industry is now worth more than a whopping $130 billion. Although games become more sophisticated constantly and new consoles and mobile devices and platforms seem to appear just as regularly, people have been playing computer games since the 1950s.[1] Although games from previous decades would not appear that impressive to today's gamers, they were in fact leaps in technology development, the first games being developed in university labs rather than game development companies we see in the business sector now.

Today, computer game development and animation is a major industry and gaming consoles or mobile devices are common in many American homes. But to see how we got to where we are today, it's interesting to look at the history of game development, and how each new invention or advancement has brought us to where we are now and where we are heading in the future of gaming.

Early Days of Computer Game Development and Animation

The very first computer game developed was the work of a British professor of computer science, Alexander Douglas, who created the first graphical computer game—a tic-tac-toe game he called OXO—in 1952 at the University of Cambridge in the UK.[2] Although quite basic by today's standards, it was an important breakthrough in computer programming research. The game—in which players played against the computer—was part of Douglas's PhD work, and intended to prove his thesis that computer/human interaction was possible.

The next advancement in computer game programming was the work of Steve Russell, a professor at the Massachusetts Institute of Technology (MIT).[3] His invention—a game called Spacewar!—was designed for the PDP-1 computer, the most sophisticated computer of its time, which mostly only universities had access to. It was the first game that was possible to play on more than one computer installation.

Although perhaps not impressive by today's standards, where the smallest of devices have more power and capability than the largest and most advanced computers of the past, technology would not have advanced as much without the innovation of computer scientists of the past. *iStock*

The very first commercial computer, called the UNIVAC I (UNIVersal Automatic Computer I), was developed by Dr. Presper Eckert and Dr. John Mauchly.[4] The men began their work in 1946, when they received a grant for $300,000 to develop the machine. After setbacks led to bankruptcy in 1950, the men were bailed out financially and finally delivered the UNIVAC I in 1951. The total cost of the machine's construction was $1 million, and only forty-six were produced for business and government use.

Progress continued in the area of computer game development, leading up to the powerful industry it is today, and the innovative technological advances that would surely boggle the minds of computer gaming's earliest pioneers.

What follows are some other key milestones in the development of computer gaming.

- *1956*: Arthur Samuel creates a checkers program on an early IBM computer, the IBM 701. By 1962, his program defeats a checkers master.
- *1961*: John Burgeson, a computer programmer, develops a baseball program on an IBM 1620 computer.
- *1964*: John Kemeny of Dartmouth University invents the BASIC programming language, with which students begin creating computer games with great enthusiasm.
- *1967–1968*: Ralph Baer invents the "Brown Box," with which the user plays tennis among other games. A year later, a company called Magnavox releases the first home video game console called Odyssey based on his prototype.
- *1972*: The first arcade game, a simple ping-pong-like game called Pong, is developed by Nolan Bushnell and Al Alcorn of Atari. Three years later, Atari releases a home version of the popular game. In 1977, the company releases the Atari 2600 home video game system, featuring a joystick, interchangeable cartridges, games in color, and various difficulty levels.
- *1980*: The infamous arcade game Pac-Man is released by Namco, and later in the year Atari releases a home version. By 1982, Ms. Pac-Man is released and becomes the best-selling arcade game of all time.

- *1985*: The Nintendo Entertainment System (NES) appears on the market, reviving an industry that was suffering at the time. In 1989, the Nintendo Game Boy makes handheld gaming all the rage, although it was not the first device of its type.
- *1991*: Sega systems introduces Sonic the Hedgehog for its Genesis device. By 1995, Sony's PlayStation—costing $100 less than the Genesis— proves to be strong competition, and by 2000 it was the leading home console in the United States.
- *2001*: Microsoft introduces its Xbox, and by 2005 is dominant in the market.
- *2006*: Dispelling of gamers sitting listless on the couch, the Nintendo Wii is released, encouraging users to move about with its motion-sensitive remotes.
- *2009*: Smart devices such as iPhones and social media apps such as Facebook make way for social games such as Angry Birds, which appeal to a broader audience than those of more traditional gamers in the past.
- *2015*: Twitch, an online video streaming service owned by Amazon, attracts 36 million viewers to its League of Legends World Championship.
- *2017*: Nintendo's Switch is released, the first mobile/home hybrid device.[5]

The Atari 2600, released in 1977, shattered records and dominated the home gaming market for thirteen years—making it the longest-running console in history. *iStock*

The Modern-Day Computer Game Development and Animation

Although this book will cover as many of these careers as possible, it will focus mostly on the following:

- *Artist/animator:* Game artists and animators have the important job of creating the environments and other animated, interactive images for computer games. They are multimedia artists and their work is enormously important to the user experience, determining how a player "moves" through the game.
- *Producer:* A computer game producer has a project-management role, putting together and overseeing a team of engineers and artists and all others involved in the creation of a game, and ensuring the project is on track.
- *Sound designer:* Every sound you hear as you play a computer game is the work of the sound designer, including hiring and recording voice actors for characters and creating the background and actions sound effects in a game.
- *Video game designer:* A video game designer—often there is more than one on a given project—is responsible for conceptualizing a new game, including the characters, story line, setting for the game, and the rules of play.
- *Video game developer:* Video game developers are the more technical side of video game creation, being computer programmers and engineers who can take a game designer's vision and perform the coding and programming to bring it to life.
- *Video game tester:* Perhaps a dream gig for any dedicated gamer, a video game tester is hired by video game companies to ensure games work as they should before being released, testing each possible action or movement in a game to ensure there are no bugs.
- *Writer:* Games that follow a certain plot or narrative require a writer to create the dialog, develop the plot, and work on character creation, just as a writer for any other medium would. The writer is part of the game designer team.

WHAT CAUSED THE 1983 VIDEO GAME CRASH

Although booming in the 1970s and into the 1980s the computer game industry plundered into recession in 1983 in what is known as the "Great North American Video Game Crash."[6] Before the crash, the industry was thriving and was worth over $2 billion—which would be worth $4 billion today.[7] But between 1983 and 1984, the industry virtually collapsed, losing 97 percent of revenue. The end of the gaming craze seemed inevitable. How did this happen?

With the onset of video game consoles, the market was simply overwhelmed by companies trying to break into what was a growing but still very niche space in the market. With literally dozens of competing console products available, consumers were confused by how to compare them all and decide which was the best and would go on to dominate the market.

In addition, each console had its own compatible games, which were very expensive (around $100) compared to today. It made sharing games with friends who had different consoles impossible. And the rush for each console manufacturer to drown the market with games to compete with other companies weakened the quality of the games available, which also had a negative impact on the industry. For example, an Atari game attempting to ride on the coattails of the popular at the time movie *E.T.* was developed at a breakneck speed of just six weeks, and failed remarkably because of its poor quality and the fact that it had so little to do with the film. Sensing the market was flailing and gaming a fading fad, some retail outlets such as toy stores began not to carry consoles.

The market was revitalized in 1985 with the release of the Nintendo Entertainment System (NES), for which fewer but much better-quality games were developed. In the end, the crash led to a shift in which companies were the leaders of the industry, as well as introducing a shifted focus on quality over quantity in games developed, an approach that regained the trust of the consumer and continues today.

The Writers Guild of America is a labor union that represents and acknowledges outstanding work with annual awards granted to writers in motion pictures, television, cable, broadcast news, and also video game writers, recognizing computer game writing as its own specialized art form.[8]

The Pros and Cons of the Computer Game Development and Animation Field

As with any career, one in computer game development and animation has upsides and downsides. But also true is that one person's pro is another person's con. If you love working in teams and collaborating with others, then this could well be the right job for you. But if you prefer to work solo, it's probably going to be frustrating for you at times. Equally if you like the rush of a hurried deadline, ever-changing specifications and due dates, and don't mind working long hours at times and weekends, then you will get a charge out of a career in computer game development and animation. But if you prefer a nine-to-five life, a predictable schedule, and don't want to bring work home with you, again, that could be a serious red flag.

Although it's one thing to read about the pros and cons of a particular career, the best way to really get a feel for what a typical day is like on the job and what the challenges and rewards are is to talk to someone who is already working in the profession, or who has in the past.

Although each profession within computer game development and animation is different, there are some generalizations that can be made when it comes to what is most challenging about the field and most gratifying.

Here are some general pros:

- You get to do what you love, be it animating, writing, testing, developing, and so on. You get to apply your technical, literary, artistic, or business skills to an industry you have a real passion for.
- The work tends to be creative and challenging.
- In this competitive field, you will have colleagues who share your passion and from whom you can learn.
- The career outlook is very good! See the next section for more.
- It is a constantly evolving field with new trends and innovations and an endless opportunity for learning.

- There's a vast degree of variety in work environments, from large corporations to start-ups to freelance work from anywhere.
- It's a field you can enter without necessarily having a college degree.
- Depending on where you work, salaries can be quite lucrative. And given that most computer game development and animation jobs are available in southern states where the cost of living is relatively lower than in cities such as San Francisco or Seattle, your earnings will stretch farther.

And here are some general cons:

- The working hours can be very long and irregular. You can expect at times to work early and late hours and also on weekends in order to meet a pressing deadline, correct a bug, adjust to a change in strategy, or any number of unpredictable issues or situations that may arise.
- For game programmers, it means a lot of time sitting at a desk. Although not all of this time is sitting at a computer doing actual coding: it also means a lot of researching, thinking, and concentrating, and largely not wanting to be interrupted.
- Because of the high level of collaboration, an artist, writer, or developer—anyone on the team—can expect to have to surrender a feature or aspect of the game they feel attached to. You have to be flexible and think as a team rather than an individual creator.
- It is a high-pressure field that requires an ability to manage stress well as well as to multitask.
- It is an extremely competitive field and breaking in and then advancing to the next level can take a lot of time, hard work, and patience.

The best part of my job is seeing a game reach the finish line. It's not an easy process, and there's a lot of stress along the way, but it's a great feeling to see something you've worked on very hard enjoyed by your fans. It doesn't always succeed, and that's not always due to lack of trying. On the contrary, you can often work above and beyond what should be expected of you, but still release something that didn't succeed.—Dan Nanni, lead designer

How Healthy Is the Job Market for Computer Gaming and Animation?

For the most part, the career outlook for the computer gaming industry is very healthy. For one thing, people will always want to play games, and technology keeps making games more sophisticated and accessible and social. This means there is plenty of room for more creative, inventive professionals in this field.

At the same time, it's a competitive field, and hard work and long hours are often required. Still, when compared with other careers in the United States, the Bureau of Labor Statistics (BLS) predicts positive growth—in some cases above the average for all careers—for the job market in the computer gaming industry. What follows are statistics published by the BLS about specific careers in the field.

ARTIST/ANIMATOR

A computer game artist/animator is a multimedia artist charged with creating all of the graphics—characters, background—for a particular game.

- *Hourly pay*: $34.87
- *Annual wage*: $72,520
- *Projected growth (2016–2026)*: 8 percent, as fast as average[9]

COMPUTER GAME PRODUCER

Computer game producers are responsible for overseeing the entire project, from compiling a team of writers, developers, designers, and so on, to ensuring that everything runs as smoothly (as possible) throughout the process and targets are reached.

- *Hourly pay*: $34.46
- *Annual wage*: $71,680
- *Projected growth (2016–2026)*: 12 percent, faster than average[10]

VIDEO GAME SOUND DESIGNER

Sound is an important part of any computer gaming experience—just try playing your favorite game with the sound turned off. The sound designer therefore has the very important function of ensuring all sounds—from character's voices to background music—enhance the experience of gameplay.

- *Hourly pay*: $20.99
- *Annual wage*: $43,660
- *Projected growth (2016–2026)*: 8 percent, as fast as average[11]

VIDEO GAME DESIGNER

A video game developer is the big-idea player, the creative mind behind the idea for a game, the person who comes up with a concept and conveys the idea to the rest of the team.

- *Hourly pay*: $50.77
- *Annual wage*: $105,590
- *Projected growth (2016–2026)*: 24 percent, much faster than average[12]

VIDEO GAME DEVELOPER

A video game developer is a programmer or engineer who brings the concept of a game to life by writing the code and ensuring the game operates as it should.

- *Hourly pay*: $40.52
- *Annual wage*: $84,280
- *Projected growth (2016–2026)*: –7 percent, decline[13]

VIDEO GAME TESTER

A video game tester has the important role of ensuring a game and all of its functions play and operate as it should, without any flaws, bugs, or unexpected crashes, errors, or other problems.

The BLS does not provide data on the job of video game tester. However, according to other sources, a game tester can expect to earn between $45,993–$74,935.[14]

WRITER

A game writer is a creative writer whose role it is to write the script of the game as well as all the dialogue and other text or voiceover that a player experiences during play.

- *Hourly pay*: $29.89
- *Annual wage*: $62,170
- *Projected growth (2016–2026)*: 8 percent, as fast as average[15]

Am I Right for a Computer Game Development and Animation Career?

This is a tough question to answer, because really the answer can only come from you. But don't despair: There are plenty of resources both online and elsewhere that can help you find the answer by guiding you through the types of questions and considerations that will bring you to your conclusion. These are covered in more detail in chapter 2. But for now, let's look at the general demands and responsibilities of a computer gaming industry career—as were mentioned previously in the section on pros and cons—and suggest some questions that may help you discover whether such a profession is a good match for your personality, interests, and the general lifestyle you want to keep in the future.

Of course no job is going to match your personality or fit your every desire, especially when you are just starting out. There are, however, some aspects to a job that may be so unappealing or simply mismatched that you may decide to opt for something else, or equally you may be so drawn to a feature of a job that any downsides are not that important.

Obviously having an ability and a passion for computer gaming, art, writing, or programming in any capacity is key to success in this field, but there are other factors to keep in mind. One way to see if you may be cut out for a career in the skilled trades is to ask yourself the following questions:

- *Would I prefer to be active and moving around during work, or would I rather mostly stay put behind a desk?* Any programmer or engineer or game tester will spend a lot of time at a desk, often in deep concentration rather than interacting with others as much as, say, a producer or a game designer or artist.
- *When something goes wrong, can I think quickly on my feet to find a solution? Do I have the leadership skills to direct others to problem solve?* Because this is a fast-paced industry where everything from game specifications to deadlines can change at any time, being flexible and staying calm under pressure is paramount. Equally important is being able to provide solutions or suggestions when something goes awry.
- *Am I a highly creative person who is also able to let an idea I may love go because others disagree or it just isn't possible?* Any creative person feels attached to his or her ideas. When you are working alone on your own vision, be it a book, a painting, a game, or what have you, you have full control over what the end result will be. It doesn't work that way when you work creatively with a team.
- *Am I able to follow directions even if I don't agree? Am I able to understand instructions quickly?* Working in a team can be a fulfilling and inspiring experience, and this field does provide that for the most part. However, it does require being comfortable with others making final decisions and understanding quickly what you are tasked with doing.
- *Can I consistently deal with people in a professional, friendly way?* Communication is a key skill to have in any profession, but particularly in computer game development and animation, because, as a designer or producer in particular it's important to convey clearly and concisely what your expectations and visions are. For any profession that relies on teamwork, no matter what your role, being able to communicate problems, potential solutions, and to understand what others are communicating makes this go far more smoothly.

TAKING A GAMBLE AS A GAME DESIGNER

Jeremy Hornik lives in Chicago and has been working in computer games for more than two decades. He has worked on trivia games, web games, educational toys, and, for the last fifteen or so years, as a designer of video slot machines. He likes to joke that artists draw, programmers program, and musicians make music . . . the person with no actual skills is the game designer.

Jeremy Hornik. *Courtesy of Jeremy Hornik*

Why did you choose to become a slot machine game designer?

I was always interested in games and, as a kid, I used to make my own board games and card games. When I got out of college, I spent a lot of my time doing theater and comedy. At some point, I got a job writing funny trivia questions for the computer game You Don't Know Jack. Working there, I kept coming up with new ideas for questions that broke the game engine, so they made me a designer. After working there, and at an educational toy company, I ended up at a slot machine company. It's been a good fit.

What is a typical day on the job for you?

I'm usually working on five or six different games at once, all at different stages. So for one game, I might be drawing pictures of a new idea, or working out some initial math ideas. For another game, I might be reviewing initial sketches with the artist. For another, it might be playing through the game and seeing if it's fun or not. And for a game that's almost ready to ship, it's paying close attention to make sure there are no tiny mistakes that might ruin the experience for a player. We often call the job "plate spinning," meaning that it's not about putting a lot of effort into one thing, it's about putting just enough effort into everything to make sure that they can all have a smooth production.

What's the best part of your job?

Working with cool, creative people. We have artists, animators, mathematicians, musicians, and engineers all working here, and they all have great ideas and interesting perspectives. Collaboration is really fun! And it's also fun to see a game of mine out in the world, where someone is playing it. It makes all the work worthwhile.

What's the worst or most challenging part of your job?

The most challenging part of my job is when my initial idea for a game, once it's been made by the cool artists and engineers and mathematicians and so forth, just isn't that much fun. Then, you have a lot of hard decisions to make. How can we change it? Is everything not much fun, or is it just one thing that's creating the problem? Will small changes make a difference, or does this need major change? Then, once I make up my mind how it needs to change, I have to ask all the other people to change everything they worked so hard on. It's not fun, but changing games to make them better is an important step . . . maybe the most important one.

What's the most surprising thing about your job?

That I've done it for eighteen years! I never thought I would find the challenges so interesting or complicated, or that the work would be so satisfying.

What kinds of qualities do you think one needs to be successful at this job?

An unending curiosity about how things work, and a willingness to be surprised. You can't know how a game will feel to play until it is up and playable. You quickly find out that the first idea is rarely the best one. Being flexible and responsive to the game that's actually in front of you (and not just chasing the game that's in your head) makes your job a lot easier and more fun.

How do you combat burnout?

There was a person who used to write monologues for the talk show host Jay Leno, which involved making the same kinds of jokes about the same kinds of people night after night after night. He said, "It's not a rut, it's a groove." You have to learn how to enjoy the incremental changes of experimentation, and the collaboration itself. The variety of tasks from day to day help too. If you can't bear sitting down at a game for one more minute, maybe it's a good time to draw for a while. If you're blocked on drawing, maybe do some flowcharting. There's always some task to distract you from other tasks.

And remember to have a life outside of your job! I do writing, and raise kids, and look at art, and still do plays every once in a while. Remember, if you're going to create broad, rich, emotional experiences, you need to have broad, rich, emotional experiences.

What would you tell a young person who is thinking about becoming a game designer?

Your ideas are important, but it's even more important to be able to explain them. Learn to talk to as many people as possible! Computer game designers need to be able to communicate between businesspeople, visually oriented people, musically oriented people, and engineers. You might be able to do one or two of those things, but usually you'll be working with other people who are better than you at them. Your job is to communicate your big idea, and to share your changes, so that everyone is working together to make the whole thing work. So you'll need to talk art to artists, algorithms to engineers, music to musicians, and profits to business people.

Summary

This chapter covered a lot of ground as far as looking more closely at the various types of professions and jobs that exist within the overarching field of computer gaming design and animation. Because the industry relies on the collaboration of so much talent from various backgrounds and skill sets, from creative writing to programming to creative conceptualization to careful, detailed testing, leaving no stone unturned. Although this chapter was not exclusively about each job specifically, most of the information is relevant to the field as a whole.

And the field of computer game development and animation is constantly evolving: In this chapter, we looked far back in history to see how the world of computer gaming has grown from simple games programmed on computers in university labs in the 1950s—computers that were far larger, less powerful, and slower than today—to the advent of game streaming and everything in between. If the past is any indication, the field and the technology on which it relies will only continue to evolve and improve.

Here are some ideas to take away with you as you move on to the next chapter:

- The field of computer game development and animation is a broad one that is ever-changing and becoming more sophisticated since it became a powerful industry more than fifty years ago.
- No day in a computer gaming industry career is the same. You can expect pleasant and less pleasant surprises, irregular schedules, long hours, and lots of activity. So it can be tiring, but never dull. And the feeling of satisfaction when a project is complete, successful, and enjoyed by millions makes it all worthwhile.

- The BLS predicts healthy growth for many roles within the computer gaming industry.
- Given all you now know about computer game development and animation professions, you may still be questioning whether such a career is right for you. This chapter provided some questions that can help you visualize yourself in real-world situations you can expect to face on the job, such as whether you see yourself collaborating in teams and working toward frequently moving targets such as schedules.

Assuming you are now more enthusiastic than ever about pursuing a career in computer game development and animation, in the next chapter we will look more closely at how you can refine your choice to a more specific job. It offers tips and advice and how to find the role and work environment that will be most satisfying to you, and what steps you can start taking—immediately!—toward reaching your future career goals.

2

Forming a Career Plan

*I*t's not easy to choose a career, yet it's one of the most important decisions you will make in your life. There are simply so many options available, and it is easy to feel overwhelmed. Particularly if you have many passions and interests, it can be hard to narrow your options down. That you are reading this book means you have decided to investigate a career in the computer gaming industry, which means you have already discovered a passion for creativity, technology, gaming, and ongoing learning. But even within the computer gaming industry, there are many choices, including what role you want to pursue, what work environment you desire, and what type of work schedule best fits your lifestyle.

Computer game development and animation fall under the same umbrella of career field, in that the creation of a computer game requires many different skill sets from many different educational backgrounds. All are needed in combination to reach the desired end result: sleek, challenging, fun, and innovative games with cutting-edge graphics, sounds, and story lines that can be played on multiple platforms. If your passion is for programming and gaming, then pursuing a career as a game developer seems like an ideal fit. If you prefer to craft engaging story lines, game writer is a good match. If you love games and strategy and the idea of inventing the next "it" game charges you up, you may be prime for a job as a game designer. And so on.

Before you can plan the path to a successful career in the computer gaming industry, it's helpful to develop an understanding of what role you want to have and in what environment you wish to work. Do you want to work in an established, multinational gaming company, or do you prefer the more entrepreneurial feel of a start-up? Or maybe you want to start your own company or design your own app game. Are you eager to relocate for a job in the gaming industry or do you prefer a freelance role from you own home?

How much education would you like to pursue? Depending on your ultimate career goal, the steps to getting there differ. Some jobs will require an associate's or bachelor's degree or higher. Others will not. To become a game tester, for example, a college degree is not required; however, to advance from this entry-level position, a degree will help you have an edge over your colleagues.[1] And of course what you choose to study or pursue certification in depends on the role you wish to pursue. A degree in computer science will not be a straight line to a job as a game artist and an art degree will not make you a programmer.

Deciding on a career means asking yourself big questions, but there are several tools and assessment tests that can help you determine what your personal strengths and aptitudes are and with which career fields and environments they best align. These tools guide you to think about important factors in choosing a career path, such as how you respond to pressure and how effectively—and how much you enjoy—you work with and communicate with people.

This chapter explores the educational requirements for various careers within the computer gaming industry, as well as options for where to go for help when planning your path to the career you want. It offers advice on how to begin preparing for your career path at any age or stage in your education, including in high school.

Planning the Plan

So where to begin? Before taking the leap and applying to college, there are other considerations and steps you can take to map out your plan for pursuing your career. Preparing your career plan begins with developing a clear understanding of what your actual career goal is.

Planning your career path means asking yourself questions that will help shape a clearer picture of what your long-term career goals are and what steps to take in order to achieve them. When considering these questions, it's important to prioritize your answers—when listing your skills, for example, put them in order of strongest to weakest. When considering questions relating to how you want to balance your career with the rest of your nonwork life, such as family and hobbies, really think about what your top priorities are and in what order.

The following are questions that are helpful to think about deeply when planning your career path.

- Think about your interests outside of the work context. How do you like to spend your free time? What inspires you? What kind of people do you like to surround yourself with, and how do you best learn? What do you really love doing?
- Brainstorm a list of the various career choices within the computer gaming industry that you are interested in pursuing. Organize the list in order of which careers you find most appealing, and then list what it is about each that attracts you. This can be anything from work environment to geographical location to the degree in which you would work with other people in a particular role.
- Research information on each job on your career choices list. You can find job descriptions, salary indications, career outlook, salary, and educational requirements information online, for example.
- Consider your personality traits. How do you respond to stress and pressure? Do you consider yourself a strong communicator? Do you work well in teams or prefer to work independently? Do you consider yourself creative? How do you respond to criticism? All of these are important to keep in mind to ensure you choose a career path that makes you happy and in which you can thrive.
- Although a career choice is obviously a huge factor in your future, it's important to consider what other factors feature in your vision of your ideal life. Think about how your career will fit in with the rest of your life, including whether you want to live in a big city or small town, how much flexibility you want in your schedule, how much autonomy you want in your work, and what your ultimate career goal is.
- The computer gaming industry is a very competitive field that, particularly when you are starting out in your career. Because it requires so much commitment, it's important to think about how willing you are to put in long hours and perform what can be very demanding work.
- While there are many lucrative careers in the computer gaming field, many job opportunities that offer experience to newcomers and recent graduates can come with relatively low salaries. What are your pay expectations, now and in the future?

Posing these questions to yourself and thinking about them deeply and answering them honestly will help make your career goals clearer and guide you in knowing which steps you will need to take to get there.

Making a decision about what kind of career to pursue can be much simpler if you ask yourself some key questions. *iStock*

Where to Go for Help

The process of deciding on and planning a career path is daunting. In many ways, the range of choices of careers available today is a wonderful thing. It allows us to refine our career goals and customize them to our own lives and personalities. In other ways, though, too much choice can be extremely daunting, and require a lot of soul-searching to navigate clearly.

Answering questions about your habits, characteristics, interests, and personality can be very challenging. Identifying and prioritizing all of your ambitions, interests, and passions can be overwhelming and complicated. It's not always easy to see ourselves objectively or see a way to achieve what matters most to us. But there are several resources and approaches to help guide you in drawing conclusions about these important questions.

- Take a career assessment test to help you answer questions about what career best suits you. There are several available online, many specifically for careers in the computer game industry.

OTHER USEFUL COURSES TO PURSUE

Although a degree in something directly related to your goal, be it engineering, programming, computer science, creative writing, or art school will definitely help you get on track to the career of your dreams, there are other courses to consider that will help you succeed once your work life is launched in the computer gaming industry.

- *Business classes.* To better understand how to succeed in a competitive and ever-changing market, to understand budgets and project management, branding, staff management, business strategy, and so on, taking business courses alongside your major will give you an edge.
- *Communication courses.* Be it internal communication with your team or company, communication with customers—be it via social media, advertising, customer support, and so on—learning to be a strong communicator is an asset in any profession.
- *Basic coding.* Whether you want to be an actual programmer or not, you will be working in a technical field. Having a basic understanding of coding and programming will help you work alongside your colleagues who are experts in the field.
- *Literature courses.* Even if you are not interested in becoming a game writer, all games rely heavily on story lines. A game designer, for example, would benefit from diving into courses on creative writing and literature.

- Consult with a career or personal coach to help you refine your understanding of your goals and how to pursue them.
- Talk with professionals working in the job you are considering and ask them what they enjoy about their work, what they find the most challenging, and what path they followed to get there.
- Educate yourself as much as possible about the industry: which games are the most popular and why, which technologies are changing both how people play and engage with games and how this affects game development and game industry business models. Stay abreast of the industry, no matter which role you wish to pursue.
- If possible, arrange to "job shadow" someone working in the field you are considering. This will enable you to experience in person what the

atmosphere is like, what a typical workday entails, how coworkers interact with each other and with management, and how well you can see yourself thriving in that role and work culture.

- Work on your portfolio. Don't wait until you can be paid professionally to start building a body of work, be it game art, game ideas, or game story lines. Hone your skills as much as you can.

Making High School Count

Once you have discovered your passion and have a fairly strong idea what type of career you want to pursue, you naturally want to start putting your career

ONLINE RESOURCES TO HELP YOU PLAN YOUR PATH

The internet is an excellent source of advice and assessment tools that can help you find and figure out how to pursue your career path. Some of these tools focus on an individual's personality and aptitude, others can help you identify and improve your skills to prepare for your career.

In addition to the sites below, you can use the internet to find a career or life coach near you—many offer their services online as well. Job sites such as LinkedIn are a good place to search for people working in a profession you'd like to learn more about, or to explore the types of jobs available in the computer gaming industry.

- At educations.com, you will find a career test designed to help you find the job of your dreams. Visit https://www.educations.com/career-test to take the test.
- Take this quiz to assess whether a career as a game designer is for you: https://www.gamedesigning.org/videogame-design-careers/.
- The Princeton Review has created a career quiz that focuses on personal interests: https://www.princetonreview.com/quiz/career-quiz.
- The Bureau of Labor Statistics provides information, including quizzes and videos, to help students up to grade 12 explore various career paths. The site also provides general information on career prospects and salaries, for example, for various jobs in the computer gaming industry and other fields. Visit https://www.bls.gov to find these resources.

path plan into motion as quickly as you can. If you are a high school student, you may feel there isn't much you can do toward achieving your career goals—other than, of course, earning good grades and graduating. But there are actually many ways you can make your high school years count toward your career in computer gaming before you have earned your high school diploma. This section will cover how you can use this period of your education and life to better prepare you for your career goal and to ensure you keep your passion alive while improving your skill set.

Even while still in high school, there are many ways you can begin working toward your career goal. *iStock*

Young adults with disabilities can face additional challenges when planning a career path. DO-IT (Disabilities, Opportunities, Internetworking, and Technology) is an organization dedicated to promoting career and education inclusion for everyone. Its website contains a wealth of information and tools to help all young people plan a career path, including self-assessment tests and career exploration questionnaires.[2]

Courses to Take in High School

Depending on your high school and what courses you have access to, there are many subjects that will help you prepare for a career in the computer gaming industry. If you go to a school that offers computer programming courses, that's a good place to start. However, there are other courses and subjects that are just as relevant to a computer gaming career. Some of them may seem unrelated initially, but they will all help you prepare yourself and develop key skills.

- *Language arts.* Because team collaboration is the essence of a job in the computer gaming industry, ensuring you know how to communicate clearly and effectively—both in spoken and written language—will be key. It helps avoid unnecessary frustration, delays, financial and time costs, and errors if you can clearly convey and understand ideas.
- *Math.* If you are aiming to be an artist in the gaming industry, you may not think math is that important. However, computers have changed how multimedia artists need to think. Need to make shapes in Illustrator using Bezier curves? You should probably take some math courses. And if you aim to own your own business, math is definitely essential for working with budgets and managing profits and losses, among other tasks.
- *Interpersonal communication/public speaking.* These courses will be an asset in any profession, including the computer gaming industry. If you need to present ideas to your team, to sales staff, to the public—these will be very important skills to hone.
- *Business and economics.* As with any type of business, if you have the ambition to run your own, knowledge gained in business and economics classes will prepare you to make smarter business and financial decisions.
- *Specialized software.* Game artists and sound engineers will need to be skilled in the technical tools they will use daily in their work. If your degree is not in multimedia art, you will need to learn these tools to succeed as a game artist.

Gaining Work Experience

The best way to learn anything is to do it. When it comes to preparing for a career in the computer gaming industry, there are several options for gaining real-world experience and getting a feel for whether you are choosing the right career for you.

The one big benefit of jobs in the computer gaming industry is you don't have to land a work-experience opportunity at an established or upcoming business to prove what you've got and what you can do. Rather than wait for someone to invite you to work for them, you are wise to keep working on your own, to show not only your talent but your passion.

This means if you want to create multimedia graphics for games, do it on your own and create a strong portfolio. If you are a strong artist in another context or medium, be sure you create works that are actually suitable to computer gaming. This also applies to game sound engineers: create a sound portfolio of game-related samples rather than a broader collection of your work.

If you want to work as a writer, it's probably not enough to create a portfolio that shows your writing chops as a short story writer or a poet. Although clearly solid writing skills and creativity are required to be a successful game writer, work on scripts that are particularly geared for games. It's an opportunity to show how you can apply your creative skills to this particular style of writing.

3-D artist Tim Simpson offers this advice for ambitious, future game artists: "Your game portfolio should contain actual game art." Although he writes that this may seem obvious, he has seen many amateur portfolios that don't. "Piles of ZBrush-only sculpts and half-finished character busts. Galleries filled with insanely detailed high poly vehicles or weapons, rendered to perfection in KeyShot," he writes. "Row upon row of awesome looking substance shader balls with displacement settings cranked through the roof. And not a single piece of realtime art to be found." His advice? Keep a final product in mind. Demonstrate an understanding of the entire workflow rather than partial.[3]

TIPS FROM THE EXPERTS: CREATING A STANDOUT
ONLINE ART PORTFOLIO

There are many online tools available—free and by subscription—that help you create a portfolio of your game artwork. Your portfolio is an important part of your application, as it allows you not only to showcase your work, but to express who you are and how passionate you are about what you do. Your résumé is important, but your portfolio is where you really show your talent and your personal style.

Game Industry Career Guide has compiled tips from successful gaming artists, as well as samples of their own portfolios. Here are some of the main takeaways:

- Always keep your demo reels and portfolio site up to date.
- Most important: Always push yourself, and get critique and feedback from the most critical person you know.
- Don't distract from the work. Make your presentation about the images, not the interface.
- The fewer clicks it takes before your gallery is presented, the better.
- Make sure your images are relevant to the job you are applying for and don't be afraid to shuffle them to fit.
- Make it easy to find you. Your contact info should be easily accessible from any point on your web page.
- Never stop creating. Revamp your site design every so often. Try to post something new as often as you can.
- Don't include anything but your *very best* work. Better ten images that rock than a "wide variety" of samples that make your work quality look "variable."[4]

Come up with game concepts if you are interested in becoming a game developer. You don't have to go so far into creating actual prototypes, but show that you can conceive of a fully thought-out, creative, engaging, and bound-to-be-a-hit game. If you're a programmer, write code for your own games—in essence, you don't need to have an official internship or job placement to gain experience you can impress potential employers with—or college entrance boards with, if you're still in high school.

It's also a good idea to arrange to job shadow with a professional in the field, in whichever capacity you find most interesting to you. This means accompanying someone to work, observing the tasks they perform, the work culture, the environment, the hours, and the intensity of the work. Talk with people you know who work in the computer gaming business

Educational Requirements

Depending on the type of job you want to pursue with the computer game industry, various levels of education are required. In some cases, it is possible to enter the field without a college degree—but in order to advance to higher levels within your career, a degree is usually preferred by employers. The following outlines the general requirements in terms of education for different computer gaming jobs.

Later in this chapter, we will discuss the considerations to keep in mind when deciding what level of education is best for you to pursue. In chapter 3, we will outline in more detail the types of programs offered and the best schools to consider, should you want to pursue post–high-school training and certification or an associate, bachelor's, or master's degree.

- *Artist/animator:* According to Study.com, this field requires no specific training to enter—meaning it may be possible to begin your career without a bachelor's degree—but for a more advanced position, employers may require a degree.[5] Game artists are expected to be proficient in tools such as Adobe imaging software such as Photoshop, Illustrator, Flash, and Maya, as well as ActionScript or FLEX.

Beyond showing off your skills, working on your own to develop games, story lines, artwork, sounds, or whatever career you want to pursue will prove your passion for this very specific industry and the way in which you want to bring your skill set to computer gaming in particular.

- *Game producer:* To become a game producer, according to Study.com, you should consider pursuing a bachelor's degree in game design, digital media, or computer science. Knowledge of computer programming languages such as Java, Visual Basic, and C++ are also required.[6]
- *Sound designer:* Educational requirements to become a sound designer range from one-year certification programs to bachelor's and master's degrees in sound design. Aspiring sound designers must be proficient in tools such as Musical Instrument Digital Interface (MIDI), as well as how to use Pro Tools, Final Cut Pro, and other sound-related software.[7]
- *Video game designer:* At minimum, a high school diploma or equivalent is required to enter the field of video game designer, but a bachelor's degree is preferred by most employers. Some knowledge of programming languages and 3-D modeling programs is also a benefit.[8]
- *Video game developer:* To become a video game developer, you will need to obtain a bachelor's degree in a computer-related field, such as computer science or software engineering.[9]
- *Video game tester:* Although it is possible to get a job as a video game tester without post–high-school or equivalent education, you will need to be knowledgeable about computer programming, hardware, and operating systems. However, to advance in your career beyond an entry-level position, an associate or bachelor's degree in a computer-related field is advised. There are also certification programs available.[10]
- *Writer:* There are always exceptions of course, particularly in creative fields, but to become a successful computer game writer, it's recommended you pursue a bachelor's degree in creative writing or script writing, to hone your storytelling skills.

Formal Training Programs: Certification Programs, Community College, and University

For some careers within the computer game industry (writer, sound designer, and game tester), a certificate program—a training program, which typically takes one or two years to complete but does not earn you a college degree—is sufficient to get an entry-level job. But a certification program is also something to consider if you think you want to pursue a degree in, say, computer game design, but you are not quite ready to commit yourself to a four-year program.

By following a certification program, you will get a foundation in game design, learn more about the profession, and get a better feel for whether it is right for you. And because networking is so important in this industry, getting to know instructors and professors and connecting with alumni of the program who are working in the field can help you establish contacts.

> You need to be open to input from up to a dozen professionals that you work with and you have to take that input and merge it into one successful project. It is not up to the producer to have the best ideas and know it all.—Christian Batist, game producer

WHY CHOOSE AN ASSOCIATE'S DEGREE?

With a two-year degree—called an associate's degree—you are qualified to apply for certain positions within the computer gaming field. Common associate's degrees offered that pertain to computer gaming are in programming, graphics, animation, artificial intelligence, and interactive simulations.

These degree programs are sufficient to give you a knowledge base to begin your career, and can serve as a basis should you decide to pursue a four-year degree later. Do keep in mind, though, that all jobs within the computer gaming industry are quite competitive. If you are prepared to put in the financial and time commitment to earn an associate's degree and are sure of the career goal you have set for yourself, consider earning a bachelor's. With so much competition out there, the more of an edge you can give yourself, the better your chances will be.

WHY CHOOSE A BACHELOR'S DEGREE?

A bachelor's degree—which usually takes four years to obtain—is a requirement for most careers related to the computer gaming industry. In general, the higher education your pursue, the better your odds are to advance in your career, which means more opportunity and often more compensation.

The difference between an associate's and a bachelor's degree is, of course, the amount of time each takes to complete. To earn a bachelor's degree, a candidate must complete forty college credits, compared with twenty for an associate's. This translates to more courses completed and a deeper exploration of degree content, even though similar content is covered in both.

Even when not required, a bachelor's degree can help advance your career, give you an edge over the competition in the field, and earn you a higher starting salary than holders of an associate's degree.

THE POWER OF SOUND

Elliot Callighan. *Courtesy of Elliot Callighan*

Elliot Callighan began playing the violin at the age of four, piano at eight, guitar at twelve, and the computer at twenty. After first pursuing architecture (and meeting his future wife), Elliot enlisted in the Army National Guard to pursue music. Five years later he had obtained multiple degrees in music composition and sound design as well as a commission as a military officer in the Army National Guard. Currently, Elliot primarily writes for games but still contributes to film, trailer, and commercial projects. He is a Soundpost Co-Chair for the Chicago Symphony Orchestra Overture Council as well as adjunct faculty in the film and game programs at DePaul University and still serves in the Army National Guard.

Why did you choose to become a composer and sound designer for games?
Music has always been the constant thing throughout my life. Tastes and interests change all the time, but I always came back to experimenting and learning more about audio and music. Originally, I hoped to be in a metal band that took over the world, but through my education and exposure to other styles, I began gravitating toward game and film soundtracks.

The second constant in my life has always been games. From Streets of Rage to Halo to World of Warcraft, I was always enthralled by these worlds, their communities, and the stories within them. Additionally, the added challenge of audio implementation gives so many opportunities for creating incredible experiences.

What is a typical day on the job for you?

I exercise at least four days a week in the morning. Having a healthy body is essential to maintaining mental dexterity, which can significantly impact your creative work and potential. After that, I'll spend a bit of time on e-mails and administrative or marketing tasks and then write or mix until lunch. This can involve just writing, recording different objects/actions, or layering existing recordings. Many times I'll grab coffee or lunch with peers or clients and finish the day essentially repeating the morning (without the exercise).

After the workday is done, I'll allow myself to actually play some games for a bit before I spend time with my wife.

What's the best part of your job?

Losing yourself in your work. When I'm working on something I'm passionate about, I'm never looking at the clock. If anything, I have the opposite problem where I'll get carried away and not notice four hours went by! Getting into that creative flow is so satisfying and fun for me.

What's the worst or most challenging part of your job?

Working with folks who either don't know what they want, or don't know how to communicate what they want. It is definitely part of my job to act as a translator in those situations. That's why it is *so* important to work on projects you're passionate about. If you aren't personally driven to make it sound exactly right for the project, a long creative/revision process can be torture.

What's the most surprising thing about your job?

How many different skills and frames of mind are necessary to be successful. To a certain extent, you have to be completely selfless, yet you have to spark inspiration in yourself to create really great outcomes. You also have to think about frequency, loudness, and personal tact very differently in different contexts that constitute this job.

What kinds of qualities do you think one needs to be successful at this job?

Perseverance, tact, and insatiable passion for what you do. As a creator, the work created can be incredibly personal. You need to be able to detach yourself from the work itself and not take criticism or differences of opinion personally. Additionally, most people will not hear the value of your work until you hit a certain threshold of professional skill. It can be very demoralizing at times—but if you want it bad enough, you'll keep going!

How do you combat burnout?

Take breaks. Take walks. Do something else for a few work hours or a day. Clearing your head can help give creative clarity once you return to your work. I will also sometimes write something purely for my own enjoyment as opposed to a project. This can help you feel that passionate drive, and be a great way to communicate your personal style to potential clients.

What would you tell a young person who is thinking about becoming a trainer?

Creative fields are very different than other disciplines. Only the exceptional are able to do it at all. It is par for the course that you can write great music, mix it, and create memorable and impactful sounds. Building your network and making friends is how you succeed. This career is a slow burn, and you need to make decisions that can financially sustain you early on.

Summary

This chapter covered a lot of ground in terms of how to break down the challenge of not only discovering what career within the computer gaming industry is right for you and in what environment, capacity, and work culture you want to work, but also how best to prepare yourself for achieving your career goal.

In this chapter, you learned about the broad range of roles that fall under the career umbrella of the computer gaming industry. And while the various careers that exist—from game designer to game tester—the chapter also pointed to many tools and methods that can help you navigate the confusing path to choosing a career that is right for you. It also addressed some of the specific training and educational options and requirements and expectations that will put you, no matter what your current education level or age, at a strong advantage in a competitive field.

Use this chapter as a guideline for how to best discover what type of career will be the right fit for you and consider what steps you can already be taking to get there. Don't forget these important tips:

- Take time to carefully consider what kind of work environment you see yourself working in, and what kind of schedule, interaction with colleagues, work culture, and responsibilities you want to have.
- Most jobs within the computer gaming industry require some specific computer skills, such as proficiency in particular programming languages and sound- and graphics-related software tools. Taking courses or getting familiar with these as early as possible—high school is not too soon!—will help you in your career or educational goals.
- Learn as much as you can about the different types of jobs available in the computer gaming field. Job-shadow a professional to get a feeling for what hours they keep, what challenges they face, and what the overall job entails. Find out what education or training they completed before launching their career.
- Investigate various colleges and certification options so you can better prepare yourself for the next step in your career path.
- Don't feel you have to wait until you graduate from high school to begin taking steps to accomplish your career goals. You can already begin by working on your portfolio so you can showcase your artwork, for example.
- Keep work/life balance in mind. The career you choose will be one of many adult decisions you make, and ensuring that you keep all of your priorities—family, location, work schedule—in mind will help you choose the right career for you, which will make you a happier person.

Chapter 3 goes into a lot more detail about pursuing the best education path. The chapter covers how to find the best value for your education and includes a discussion about financial aid and scholarships. At the end of chapter 3, you should have a much clearer view of the educational landscape and how and where you fit in.

3

Pursuing the Education Path

*M*aking decisions about your education path can be just as daunting as choosing a career path. It is a decision that demands understanding not only what kind of education or training is required for the career you want, but also what kind of school or college you want to attend. There is a lot to consider no matter what area of study you want to pursue, and depending on the type of job you want to have within the computer gaming and animation field.

Now that you've gotten an overview of the different degree and certificate options that can prepare you for your future career, this chapter will dig more deeply into how to best choose the right type of study for you. Even if you are years away from earning your high school diploma or equivalent, it's never too soon to start weighing your options, thinking about the application process, and of course taking time to really consider what kind of educational track and environment will suit you best.

Some people choose to start their careers right away after graduating with a high school degree or equivalent. For some jobs, as mentioned—such as game tester—this can be sufficient providing you have the computer-related knowledge to do the job well. Not everyone wants to take time to go to college or pursue other forms of academic-based training. But if you are interested in following the education path—from earning a certificate in game design or multimedia art to a four-year university degree—this chapter will help you navigate the process of deciding on the type of institution you would most thrive in, determining what type of degree you want to earn, and looking into costs and how to find help in meeting them.

The chapter will also give you advice on the application process, how to prepare for any entrance exams such as the SAT or ACT that you may need to take, and how to communicate your passion, ambition, and personal experience in a personal statement. When you've completed this chapter, you should have a good sense of what kind of post–high-school education is right for you and how to ensure you have the best chance of being accepted at the institution of your choice.

Finding a Program or School That Fits Your Personality

Before we get into the details of good schools for each profession, it's a good idea for you to take some time to consider what "type" of school will be best for you. Just as with your future work environment, understanding how you best learn, what type of atmosphere best fits your personality, and how and where you are most likely to succeed will play a major part in how happy you will be with your choice. This section will provide some thinking points to help you refine what kind of school or program is the best fit for you.

Note this list does not assume you intend to attend a four-year college program or complete a certification program—some of the questions may therefore be more relevant to you, depending on the path of study you mean to follow.

If nothing else, answering questions like the following ones can help you narrow your search and focus on a smaller sampling of choices. Write your answers to these questions down somewhere where you can refer to them often, such as in your notes app on your phone:

- *Size:* Does the size of the school matter to you? Colleges and universities range from sizes of five hundred or fewer students to twenty-five thousand students. If you are considering college or university, think about what size of class you would like, and what the right instructor-to-student ratio is for you.
- *Community location:* Would you prefer to be in a rural area, a small town, a suburban area, or a large city? How important is the location of the school in the larger world to you? Is the flexibility of an online degree or certification program attractive to you, or do you prefer more on-site, hands-on instruction?
- *Length of study:* How many months or years do you want to put into your education before you start working professionally?
- *Housing options:* If applicable, what kind of housing would you prefer? Dorms, off-campus apartments, and private homes are all common options.

- *Student body:* How would you like the student body to look? Think about coed versus all-male and all-female settings, as well as the makeup of minorities, how many students are part-time versus full-time, and the percentage of commuter students.
- *Academic environment:* Consider which majors are offered and at which levels of degree. Research the student-faculty ratio. Are the classes taught often by actual professors or more often by the teaching assistants? Find out how many internships the school typically provides to students. Are independent study or study abroad programs available in your area of interest?
- *Financial aid availability/cost:* Does the school provide ample opportunities for scholarships, grants, work-study programs, and the like? Does cost play a role in your options? (For most people, it does.)
- *Support services:* Investigate the strength of the academic and career placement counseling services of the school.
- *Social activities and athletics:* Does the school offer clubs that you are interested in? Which sports are offered? Are scholarships available?
- *Specialize programs:* Does the school offer honors programs or programs for veterans or students with disabilities or special needs?

Not all of these questions are going to be important to you and that's fine. Be sure to make note of aspects that don't matter so much to you too, such as size or location. You might change your mind as you go to visit colleges, but it's important to make note of where you're at to begin with.

U.S. News & World Report puts it best when it reports that the college that fits you best is one that:

- Offers a degree that matches your interests and needs
- Provides a style of instruction that matches the way you like to learn
- Provides a level of academic rigor to match your aptitude and preparation
- Offers a community that feels like home to you
- Values you for what you do well[1]

MAKE THE MOST OF CAMPUS VISITS

If it's at all practical and feasible, you should visit the campuses of all the schools you're considering. To get a real feel for any college or university, you need to walk around the campus, spend some time in the common areas where students hang out, and sit in on a few classes. You can also sign up for campus tours, which are typically given by current students. This is another good way to see the campus and ask questions of someone who knows. Be sure to visit the specific school/building that covers your possible major as well. The website and brochures won't be able to convey that intangible feeling you'll get from a visit.

In addition to the questions listed in the section "Finding a College That Fits Your Personality," consider these questions as well. Make a list of questions that are important to you before you visit.

- What is the makeup of the current freshman class? Is the campus diverse?
- What is the meal plan like? What are the food options?
- Where do most of the students hang out between classes? (Be sure to visit this area.)
- How long does it take to walk from one end of the campus to the other?
- What types of transportation are available for students? Does campus security provide escorts to cars, dorms, etc., at night?

In order to be ready for your visit and make the most of it, consider these tips and words of advice:

- Be sure to do some research. At the least, spend some time on the college website. Make sure your questions aren't addressed adequately there first.
- Make a list of questions.
- Arrange to meet with a professor in your area of interest or to visit the specific school.
- Be prepared to answer questions about yourself and why you are interested in this school.
- Dress in neat, clean, and casual clothes. Avoid overly wrinkled clothing or anything with stains.
- Listen and take notes.
- Don't interrupt.

- Be positive and energetic.
- Make eye contact when someone speaks directly to you.
- Ask questions.
- Thank people for their time.

Finally, be sure to send thank-you notes or e-mails after the visit is over. Remind recipients when you visited the campus and thank them for their time.

Hopefully, this section has impressed upon you the importance of finding the right fit for your chosen learning institution. Take some time to paint a mental picture about the kind of university or school setting that will best complement your needs. Then read on for specifics about each degree.

In the academic world, accreditation matters, and is something you should consider when choosing a school. Accreditation is basically a seal of approval that schools promote to let prospective students feel sure the institution will provide a quality education that is worth the investment and will help graduates reach their career goals. Future employers will want to see that the program you completed has such a seal of quality, so it's something to keep in mind when choosing a school.

Determining Your Education Plan

There are many options, as mentioned, when it comes to pursing an education in the computer gaming and animation field. These include vocational schools, two-year community colleges, and four-year colleges. This section will help you select the track that is most suited to you.

Whether you are opting for a certificate program, a two-year, or four-year degree, you will find there are many choices of institutes and schools offering a variety of programs at different costs and durations (in the case of certificate programs, twelve to eighteen months is usually the average for full-time participants to complete the required course load). Because of this, it is important to narrow down the options and compare them closely.

CONSIDERING A GAP YEAR

Taking a year off between high school and college, often called a gap year, is normal, perfectly acceptable, and almost required in many countries around the world, and it is becoming increasingly acceptable in the United States as well. Because the cost of college has gone up dramatically, it literally pays for you to know going in what you want to study, and a gap year—well spent—can do lots to help you answer that question. It can also give you an opportunity to explore different types of computer-gaming-related jobs to help you find a deeper sense of what you'd like to study when your gap year has ended.

Some great ways to spend your gap year include joining the Peace Corps or other organizations that offer opportunities for work experience. But even if the experience has nothing to do directly with computers or games, a gap year can help you see things from a new perspective. Consider enrolling in a mountaineering program or other gap year-styled program, backpacking across Europe or other countries on the cheap (be safe and bring a friend), find a volunteer organization that furthers a cause you believe in or that complements your career aspirations, join a Road Scholar program (see https://www.roadscholar.org/), teach English in another country (see https://www.gooverseas.com/blog/best-countries-for-seniors-to-teach-english -abroad for more information), or work and earn money for college!

Many students will find that they get much more out of college when they have a year to mature and to experience the real world. The American Gap Year Association reports from their alumni surveys that students who take gap years show improved civic engagement, improved college graduation rates, and improved GPAs in college.

See the association's website at https://gapyearassociation.org/ for lots of advice and resources if you're considering a potentially life-altering experience.

It's a good idea to select roughly five to ten schools in a realistic location (for you) that offer the degree or certification you want to earn. If you are considering online programs, include these in your list. Of course, not every school near you or that you have an initial interest in will probably grant the degree you want of course, so narrow your choices accordingly. With that said, consider attending a university in your resident state, if possible, which will save you lots of money if you attend a state school. Private institutions don't typically discount resident student tuition costs.

Be sure you research the basic GPA and SAT or ACT requirements of each school as well. Although some community colleges do not require standardized tests for the application process, others do.

If you are planning to apply to a college or program that requires the ACT or SAT, advisers recommend that students take both the ACT and the SAT tests during their junior year (spring at the latest). You can retake these tests and use your highest score, so be sure to leave time to retake early in your senior year if needed. You want your best score to be available to all the schools you're applying to by January of your senior year, which will also enable them to be considered with any scholarship applications. Keep in mind these are general timelines—be sure to check the exact deadlines and calendars of the schools to which you're applying!

Once you have found five to ten schools in a realistic location for you that offer the degree or certification in question, spend some time on their websites studying the requirements for admissions. Important factors weighing on your decision of what schools to apply to should include whether or not you meet the requirements, your chances of getting in (but shoot high!), tuition costs and availability of scholarships and grants, location, and the school's reputation and licensure/graduation rates.

Most colleges and universities will list the average stats for the last class accepted to the program, which will give you a sense of your chances of acceptance.

The order of these characteristics will depend on your grades and test scores, your financial resources, work experience, and other personal factors. Taking everything into account, you should be able to narrow your list down to the institutes or schools that best match your educational or professional goals as well as your resources and other factors such as location and duration of study.

Schools to Consider When Pursuing a Career in Computer Game Development and Animation

Some schools and programs have stronger reputations than others. Although you can certainly have a successful and satisfying career and experience without going to the number one school in your field of study, it is a good idea to shop around, to compare different schools and get a sense of what they offer and what features of each are the most important—or least—to you.

Keep in mind that what is great for one person may not be as great for someone else. What might be a perfect school for you might be too difficult, too expensive, or not rigorous enough for someone else. Keep in mind the advice of the previous sections when deciding what you really need in a school.

Also, bear in mind there are different degree programs you can follow in order to end up with the same career—computer game testers and developers, for example, all require a background in computer science. So while programs designed specifically for computer gaming careers exist, you don't necessarily have to attend one to prepare for a future in the field.

GREAT SCHOOLS FOR GAME DESIGN

The Princeton Review has compiled a list of the top fifty schools for game design.[2] The list is based on institutional survey data, including academic offerings in game design, faculty credentials, and career outcomes. The top ten programs, according to this research, are listed here in order:

1. University of Southern California, Los Angeles, California
2. New York University, New York, New York
3. Becker College, Worcester, Massachusetts
4. DigiPen Institute of Technology, Redmond, Washington
5. Hampshire College, Amherst, Massachusetts
6. Drexel University, Philadelphia, Pennsylvania
7. Michigan State University, East Lansing, Michigan
8. Rochester Institute of Technology, Rochester, New York
9. Worcester Polytechnic Institute, Worcester, Massachusetts
10. University of Utah, Salt Lake City, Utah

GREAT SCHOOLS FOR DIGITAL ANIMATION

If you want to pursue a career in computer game animation, consider these schools, ranked by the website College Rank, which states on its site that these animation programs "have graduates with a high rate of job placement, and offer the most up-to-date technology and offer various types of animation training, including 2-D, 3-D, lighting, storyboarding and scriptwriting. Another factor in choosing colleges for this list is that they must be accredited."[3]

Here are the top ten programs, listed in order:

1. Rhode Island School of Design, Providence, Rhode Island
2. Ringling College of Art and Design, Sarasota, Florida
3. California Institute of the Arts, Valencia, California
4. University of Southern California, Los Angeles, California
5. Savannah College of Art and Design, Savannah, Georgia
6. Carnegie Mellon University, Pittsburgh, Pennsylvania
7. Rochester Institute of Technology, Rochester, New York
8. University of California, Los Angeles, Los Angeles, California
9. University of Central Florida, Orlando, Florida
10. Maryland Institute College of Art, Baltimore, Maryland

GREAT SCHOOLS FOR COMPUTER SCIENCE

If you are interested in a career as a game developer, designer, or tester, a computer science degree will give you the technical background and knowledge to write code (or test it) to create exciting games that behave as they should. *U.S. News & World Report* has compiled a list of the best computer science undergrad programs in the United States, based on a survey of academics at peer institutions.[4] The following lists the top ten, in order:

1. Carnegie Mellon University, Pittsburgh, Pennsylvania (tie for first)
2. Massachusetts Institute of Technology, Cambridge, Massachusetts (tie for first)
3. Stanford University, Stanford, California (tie for first)
4. University of California—Berkeley, Berkeley, California (tie for first)
5. University of Illinois—Urbana-Champaign, Urbana, Illinois
6. Cornell University, Ithaca, New York (tie for sixth)

7. University of Washington, Seattle, Washington (tie for sixth)
8. Georgia Institute of Technology, Atlanta, Georgia (tie for eighth)
9. Princeton University, Princeton, New Jersey (tie for eighth)
10. University of Texas—Austin, Austin, Texas

GREAT SCHOOLS FOR CREATIVE WRITING

If you are eager to launch your career as a game writer, you should consider earning a degree that will hone your creative writing skills and understanding of the elements of good narratives. There are many master of fine arts programs in the United States that focus on creative writing, including fiction or script writing, but as an undergrad you might consider an English literature degree and take as many creative writing courses as you can along the way. Or, you might choose to focus more intently on creative writing, by attending one of the following schools offering undergraduate (bachelor's) programs in creative writing.

This list has been compiled by the Koppelman Group:[5]

1. Columbia University, New York, New York
2. Emory University, Atlanta, Georgia
3. Washington University in St. Louis, St. Louis, Missouri
4. Princeton University, Princeton, New Jersey
5. Middlebury College, Middlebury, Vermont
6. Emerson College, Boston, Massachusetts
7. Cornell University, Ithaca, New York
8. Hamilton College, Clinton, New York
9. Bucknell University, Lewisburg, Pennsylvania
10. Kenyon College, Gambier, Ohio

What's It Going to Cost You?

So, the bottom line—what will your education end up costing you? Of course, that depends on many factors, including the type and length of degree or certification, where you attend (in-state or not, private or public institution), how much in scholarships or financial aid you're able to obtain, your family or personal income, and many other factors.

Generally speaking, there is about a 3 percent annual increase in tuition and associated costs to attend college. In other words, if you are expecting to attend college two years after this data was collected, you need to add approximately 6 percent to these numbers. Keep in mind this assumes there is no financial aid or scholarships of any kind.

School can be an expensive investment, but there are many ways to seek help paying for your education. *iStock*

WRITING A GREAT PERSONAL STATEMENT FOR ADMISSION

The personal statement you include with your application to college is extremely important, especially when your GPA and SAT/ACT scores are on the border of what is typically accepted. Write something that is thoughtful and conveys your understanding of the profession you are interested in, as well as your desire to practice in this field. Why are you uniquely qualified? Why are you a good fit for this university? These essays should be highly personal (the "personal" in personal statement). Will the admissions professionals who read it, along with hundreds of others, come away with a snapshot of who you really are and what you are passionate about?

Look online for some examples of good ones, which will give you a feel for what works. Be sure to check your specific school for length guidelines, format requirements, and any other guidelines they expect you to follow.

And of course, be sure to proofread it several times and ask a professional (such as your school writing center or your local library services) to proofread it as well.

Financial Aid: Finding Money for Education

Finding the money to attend college, whether is two or four years, an online program, or a vocational career college, can seem overwhelming. But you can do it if you have a plan before you actually start applying to college. If you get into your top-choice university, don't let the sticker cost turn you away. Financial aid can come from many different sources and it's available to cover all different kinds of costs you'll encounter during your years in college, including tuition, fees, books, housing, and food.

The good news is that universities more often offer incentive or tuition discount aid to encourage students to attend. The market is often more competitive in the favor of the student and colleges and universities are responding by offering more generous aid packages to a wider range of students than they used to. Here are some basic tips and pointers about the financial aid process:

- You apply for financial aid during your senior year. You must fill out the Free Application for Federal Student Aid (FAFSA) form, which can be filed starting October 1 of your senior year until June of the year you graduate.[6] Because the amount of available aid is limited, it's best to apply as soon as you possibly can. See https://studentaid.ed.gov/sa/fafsa to get started.

- Be sure to compare and contrast deals you get at different schools. There is room to negotiate with universities. The first offer for aid may not be the best you'll get.
- Wait until you receive all offers from your top schools and then use this information to negotiate with your top choice to see if they will match or beat the best aid package you received.
- To be eligible to keep and maintain your financial aid package, you must meet certain grade/GPA requirements. Be sure you are very clear on these academic expectations and keep up with them.
- You must reapply for federal aid every year.

Watch out for scholarship scams! You should never be asked to pay to submit the FAFSA form ("free" is in its name) or be required to pay a lot to find appropriate aid and scholarships. These are free services. If an organization promises you you'll get aid or that you have to "act now or miss out," these are both warning signs of a less reputable organization.

Also, be careful with your personal information to avoid identity theft as well. Simple things like closing and exiting your browser after visiting sites where you entered personal information goes a long way. Don't share your student aid ID number with anyone either.

It's important to understand the different forms of financial aid that are available to you. That way, you'll know how to apply for different kinds and get the best financial aid package that fits your needs and strengths. The two main categories that financial aid falls under is gift aid, which don't have to be repaid, and self-help aid, which are either loans that must be repaid or work-study funds that are earned. The next sections cover the various types of financial aid that fit in one of these areas.

GRANTS

Grants typically are awarded to students who have financial needs but can also be used in the areas of athletics, academics, demographics, veteran support, and special talents. They do not have to be paid back. Grants can come from federal agencies, state agencies, specific universities, and private organizations. Most federal and state grants are based on financial need.

Examples of grants are the Pell Grant, SMART Grant, and the Federal Supplemental Educational Opportunity Grant (FSEOG). Visit the US Department of Education's Federal Student Aid site for lots of current information about grants (see https://studentaid.ed.gov/types/grants-scholarships).

SCHOLARSHIPS

Scholarships are merit-based aid that does not have to be paid back. They are typically awarded based on academic excellence or some other special talent, such as music or art. Scholarships also fall under the areas of athletic-based, minority-based, aid for women, and so forth. These are typically not awarded by federal or state governments, but instead come from the specific university you applied to as well as private and nonprofit organizations.

Be sure to reach out directly to the financial aid officers of the schools you want to attend. These people are great contacts that can lead you to many more sources of scholarships and financial aid. Visit http://www.gocollege.com/financial-aid/scholarships/types/ for lots more information about how scholarships in general work.

LOANS

Many types of loans are available especially to students to pay for their post-secondary education. However, the important thing to remember here is that loans must be paid back, with interest. Be sure you understand the interest rate you will be charged. This is the extra cost of borrowing the money and is usually a percentage of the amount you borrow. Is this fixed or will it change over time? Is the loan and interest deferred until you graduate (meaning you don't have to begin paying it off until after you graduate)? Is the loan subsidized (meaning the federal government pays the interest until you graduate)?

These are all points you need to be clear about before you sign on the dotted line.

There are many types of loans offered to students, including need-based loans, non-need-based loans, state loans, and private loans. Two very reputable federal loans are the Perkins Loan and the Direct Stafford Loan. For more information about student loans, start at https://bigfuture.collegeboard.org /pay-for-college/loans/types-of-college-loans.

FEDERAL WORK-STUDY

The US federal work-study program provides part-time jobs for undergraduate and graduate students with financial need so they can earn money to pay for educational expenses. The focus of such work is on community service work and work related to a student's course of study. Not all colleges and universities participate in this program, so be sure to check with the school financial aid office if this is something you are counting on. The sooner you apply, the more likely you will get the job you desire and be able to benefit from the program, as funds are limited. See https://studentaid.ed.gov/sa/types/work-study for more information about this opportunity.

GAMES FOR LEARNING

Christian Batist. *Courtesy of Christian Batist*

Christian Batist, born and raised in the Netherlands, started a web design company while still in university in 1995. He sold this company to become an internet publisher at the largest media company in the Netherlands, where he introduced world-leading teen virtual world Habbo Hotel. Two years later, he stepped up to become SVP of Marketing at Sulake Corporation (Habbo), leading a team of more than thirty professionals. In 2010 he started a game consultancy business, with sixty-plus clients to date (such as Stardoll, Spil Games, Nickelodeon, WildWorks, and Glu Mobile).

In 2012 Christian founded Perfect Earth, with the aim to put his network and experience to work toward a better world for future generations, developing a portfolio of mobile and traditional games that allow children to discover the impact of their actions on the environment.

Why did you choose to become a game producer?

That was never a conscious choice, I am afraid. When I was in school, this was not really an available option. . . . These days we have over forty schools in the Netherlands teaching some form of game design and/or game development. But you don't really become a producer straight out of school even today. I spent half my childhood playing board games, and when you passionately like playing games, you will have ideas on how to improve on them. If you then start sharing those ideas with the makers of such games and they resonate well with their ideas on where to take their games, you simply just find yourself landing a game design gig. And it simply continues from there as long as you deliver viable design concepts. But I spent a decade of my professional life being sidetracked on website publishing, before I got involved with my first game and even then it was not as a producer but as the country manager of the worldwide virtual world phenomenon Habbo, responsible for marketing and sales, rather than game design. As I was trying to improve the financial results of this game, when I became senior vice president of marketing, I also got to have influence on the future roadmap of the game, and this is where I finally got through to the core of game design. Having worked on this massive title, with 20 million unique monthly players, I earned my ranks to work with other great titles, such as Animal Jam and Design Home.

What is a typical day on the job for you?

Oh man, that's the hardest question to answer. I do not really have typical days, as I work for my own company as well as for clients. But on any given day I might find myself writing narrative, documenting gameplay mechanics in the greatest detail, balancing a virtual economy, plotting new price point experiments, or writing instructions for a user interface artist. And as the producer you have P&L responsibility, so I may also spend some time reporting to management, hiring or reviewing employees, planning the roadmap with a third-party development studio, explaining to investors what their money was used for and watching playtest videos of dozens of players to see if our clever designs stand the test of reality.

What's the best part of your job?

Working with small teams of creative professionals that are passionate about what they do. Every day we learn and every day we have fun. This is hardly a job in the traditional sense of the word. It is a calling and a lifestyle and we are ever so fortunate to have found what makes us tick. Just imagine the look on my son's face, at the age of six, when he inquired what I actually do at work and I told him "I make games." Priceless!

What's the worst or most challenging part of your job?

The most challenging part of my job is to get the game delivered on time and within budget. You would think that that gets easier with experience, but as the market is constantly changing, both in terms of technology and in terms of trends in consumer behavior, it remains truly difficult. Producers of a mobile game may get six to twelve months to deliver a commercially viable product. That includes a period of development and a period of soft launch, during which the game is actually published in a limited territory to test if the performance of the game is on target with the original objectives. The trick is to get a so-called minimum viable product out as soon as possible so you can measure its performance and start to iterate on the design to improve it further. But the budget for the new title is mostly fixed, as nobody likes to sign a blank check. And depending on the planned marketing efforts, the deadline may not be very flexible, so as the project nears the deadline, or the budget limit, teams will need to give it their all to make it work. And it certainly does not always work out as planned. I have to ask for more time and/or more budget more often than not.

What's the most surprising thing about your job?

The most surprising thing about my job is that the eventual end user might experience our game completely different from what we envisioned. When the play test videos come in, we hold our breath in anticipation. These are ruthless videos, combining visual reports on how players navigate through our game with their comments in full-on audio. Until that moment we have tested the game in-house and with friends from the industry. All of us are on the latest devices and with high speed internet connections. But the play tests are performed by "real people" in real-life situations, often on legacy phones with poor connections. A game that takes me three seconds to download, may take somebody in Midwestern United States two minutes. That's eternity, and some people already drop out before having seen any of the awesome art and magnificent mechanics we have created for them. That's cruel, but that's reality. We spend twelve months and hundreds of thousands of dollars on a game they can download for free and they get to tell us it sucks within minutes into the game. But it is not always this bad. We also get great insights that help us improve the game and new iterations will show great increases in player retention, so we are happy to go through the necessary pain.

What kinds of qualities do you think one needs to be successful at this job?

You need to be open to input from up to a dozen professionals that you work with and you have to take that input and merge it into one successful project. It is not up to the producer to have the best ideas and know it all. Like any leader of any team you have to make sure you create an environment that welcomes valuable input. Your team needs to feel safe to voice their ideas and concerns about the project. All you do is provide a vision and facilitate all team members to be able to give it their best. Oh, and budgeting is not a luxury skill to have. Going over budget, or delivering late, is not the best way to land your next project, after all.

How do you combat burnout?

I doubt if there is much burnout in this industry. We work hard, but we also play hard. I know that a lot of hardworking people will use this term loosely, but the game industry takes that playful mentality super serious. We have plenty of team outings, most of which involve some sort of play and all of which are aimed at opening our minds and stimulating creativity. But the best way to prevent any stress as a producer is to make sure the plan is realistic from the start and all parties involved understand their role in the whole of the production to the last detail. Scrum tools and frequent short status updates help with that. Fostering transparency and open communication can help prevent individuals within the team from drowning in their workload without others noticing. Cultivating a good sense of joint effort versus joint reward will guarantee that colleagues will help each other when needed. And as the producer you simply have to care about all aspects of the product, from the app icon in the App Store, to the speedy response of your player support team. There's no hiding and no shying away from that.

What would you tell a young person who is thinking about becoming a game producer?

Be humble. Play lots of games. Try and document the core loops of the games and how the different mechanics balance the overall gameplay. Then try and write down in the greatest detail how you would perfect these games. What's missing? What's frustrating? Where did they lose interest to play further? To become a great producer, you first need to be a great designer. Put in the time. Spend five to ten years designing games, as part of a small team, before you accept the responsibility of game producer. And please, do not apply for a game producer role if you do not have this experience, because in the best-case scenario we will laugh at your ignorance, but we may actually be very offended by your lack of respect of the role in question.

Summary

This chapter covered all the aspects of college and postsecondary schooling/ certification that you'll want to consider as you move forward. Remember that finding the right fit is especially important, as it increases the chances that you'll stay in school and earn your degree or certificate, as well as have an amazing experience while you're at it. The careers covered in this book have varying educational requirements, which means that finding the right school or program can be very different depending on your career aspirations.

In this chapter, we discussed how to evaluate and compare your options in order to get the best education for the best deal. You also learned a little about scholarships and financial aid, how the SAT and ACT tests work, if applicable, and how to write a unique personal statement that eloquently expresses your passions.

Use this chapter as a jumping off point to dig deeper into your particular area of interest, but don't forget these important points:

- Take the SAT and ACT tests early in your junior year so you have time to take them again. Most universities automatically accept the highest scores.
- Make sure that the institution you plan to attend has an accredited program in your field of study. Some professions follow national accreditation policies, while others are state-mandated and therefore differ across state lines. Do your research and understand the differences.
- Don't underestimate how important campus visits are, especially in the pursuit of finding the right academic fit. Come prepared to ask questions not addressed on the school website or in the literature.
- Your personal statement is a very important piece of your application that can set you apart from others. Take the time and energy needed to make it unique and compelling.
- Don't assume you can't afford a school based on the sticker price. Many schools offer great scholarships and aid to qualified students. It doesn't hurt to apply. This advice especially applies to minorities, veterans, and students with disabilities.
- Don't lose sight of the fact that it's important to pursue a career that you enjoy, are good at, and are passionate about! You'll be a happier person if you do so.

At this point, your career goals and aspirations should be jelling. At the least, you should have a plan for finding out more information. Remember to do the research about the university, school, or degree/certificate program before you reach out and especially before you visit. Faculty and staff find students who ask challenging questions much more impressive than those who ask questions that can be answered by spending ten minutes on the school website.

Chapter 4 goes into detail about the next steps—writing a résumé and cover letter, interviewing well, follow-up communications, and more. This information is not just for college grads; you can use it to secure internships, volunteer positions, summer jobs, and other opportunities. In fact, the sooner you can hone these communication skills, the better off you'll be in the professional world.

4

Writing Your Résumé and Interviewing

You are now well on your way to mapping your path to achieve your career goals in the computer gaming and animation field. With each chapter of this book, we have narrowed the process from the broadest of strokes—what is the computer gaming industry, and what kinds of jobs exist in it—to how to plan your strategy and educational approach to making your dream job a reality.

This chapter will cover the steps involved in applying for jobs or schools: how to prepare an effective résumé and slam dunk an interview. Your résumé is your opportunity to summarize your experience, training, education, and goals and attract employers or school administrators. The goal of the résumé is to land the interview, and the goal of the interview is to land the job. Even if you do not have much working experience, you can still put together a résumé that expresses your interests and goals and the activities that illustrate your competence and interest.

As well as a résumé, you will be expected to write a cover letter that is basically your opportunity to reveal a little bit more about your passion, your motivation for a particular job or educational opportunity, and often to express more about you personally to give a potential employer a sense of who you are and what drives you. And particularly because you are striving for a career in a very competitive and creative field, it's wise to ensure your uniqueness, creative flair, and passion for gaming, writing, animation, or programming—whatever your goal—comes through.

Giving the right impression is undoubtedly important, but don't let that make you nervous. In a résumé, cover letter, or interview, you want to put forward your best but your genuine self. Dress professionally, proofread carefully, but ensure you are being yourself. In this chapter, we will cover all of these important aspects of the job-hunting process and by the end you will feel confident and ready to present yourself as a candidate for the job you really want.

When looking for samples of résumé content and design, be as specific as possible to the type of job you want. For example, the online Game Industry Career Guide offers samples of résumés of candidates who sought and landed a job in the computer gaming field.[1]

Writing Your Résumé

Writing your first résumé can feel very challenging because you have likely not yet gained a lot of experience in a professional setting. But don't fret: employers understand that you are new to the workforce or to the particular career you are seeking. The right approach is never to exaggerate or invent experience or accomplishments, but to present yourself as someone with a good work ethic, a genuine interest in the particular job or company, and use what you can to present yourself authentically and honestly.

There are some standard elements to an effective résumé that you should be sure to include. At the top should be your name, of course, as well as e-mail address or other contact information. Always list your experience in chronological order, beginning with your current or most recent position—or whatever experience you want to share. If you are a recent graduate with little work experience, begin with your education. If you've been in the working world for a while, you can opt to list your education or any certification you have at the end. The important thing is to present the most important and relevant information at the top. With only six seconds to make an impression, your résumé needs to be easy to navigate and read.

Before you even begin to write your résumé, do your research. Make sure you get a good sense of what kind of candidate or applicant a school or an employer is looking for. You want to not only come across as competent and qualified but also to seem like just the right fit for just that job within that organization.

Once you know more about the intended audience—organization, institution, or individual—of your résumé, you can begin to make a list of all the relevant experience and education you have. You may need to customize your résumé for different purposes to ensure you are not filling it with information that does not directly link to your qualifications for a particular job.

The *Washington Post* reports that mentioning a passion for video games for any profession, not just one in the actual industry, gives candidates an edge. The reason? A background in either creating or playing video games helps potential employees with online collaboration and problem solving, among other skills.[2]

Highlight your education where you can—any courses you've taken be it in high school or through a community college or any other place that offers training related to your job target. Also highlight any hobbies or volunteer experience you have—again, only as it relates to the job you are after.

Remember to have a life outside of your job! I do writing, and raise kids, and look at art, and still do plays every once in a while. Remember, if you're going to create broad, rich, emotional experiences, you need to have broad, rich, emotional experiences.—Jeremy Hornik, game designer

Your résumé is a document that sums up who you are and indicates in what ways you will be an asset to your future employer. But the trick is it should also be concise: one page is usually appropriate, especially for your very first résumé.

Before preparing your résumé, try to connect with a hiring professional—a human resources person or hiring manager—in a similar position or organization you are interested in. He or she can give you advice on what employees look for and what information to highlight on your résumé, as well as what types of interview questions you can expect.

As important as your résumé's content is the way you design and format it. You can find several samples online of résumés that you can be inspired by. At The Balance Careers, for example, you can find many templates and design ideas.[3] You want your résumé to be attractive to the eye and formatted in a way that makes the key points easy to spot and digest—according to some research, employees take an average of six seconds to review a résumé, so you don't have a lot of time to get across your experience and value.

Your headline will appear just below your name and should summarize—in 120 characters—who you are, what you do, what you are interested in doing, and what you are motivated by. Take your time with this—it is your opportunity to sell yourself in a brief and impactful manner. Related but separate is

LINKING IN WITH IMPACT

As well as your paper or electronic résumé, creating a LinkedIn profile is a good way to highlight your experience and promote yourself, as well as to network. Joining professional organizations or connecting with other people in your desired field are good ways to keep abreast of changes and trends and work opportunities.

The key elements of a LinkedIn profile are your photo, your headline, and your profile summary. These are the most revealing parts of the profile and the ones employers and connections will base their impression of you on.

The photo should be carefully chosen. Remember that LinkedIn is not Facebook or Instagram: it is not the place to share a photo of you acting too casually on vacation or at a party. According to Joshua Waldman, author of *Job Searching with Social Media for Dummies*, the choice of photo should be taken seriously and be done right. His tips:

- Choose a photo in which you have a nice smile.
- Dress in professional clothing.
- Ensure the background of the photo is pleasing to the eye. According to Waldman, some colors—like green and blue—convey a feeling of trust and stability.
- Remember it's not a mug shot. You can be creative with the angle of your photo rather than stare directly into the camera.
- Use your photo to convey some aspect of your personality.
- Focus on your face. Remember visitors to your profile will see only a small thumbnail image, so be sure your face takes up most of it.[4]

your summary section. Here, you can share a little more about yourself than in your headline, but it should still be brief. Walman recommends your summary take no more than thirty seconds to read aloud (so yes, time yourself!); that it be short (between five and ten lines or three to five sentences), concise, and unique; and that it tells a story.

> If you are applying for a job as a writer or artist in particular, it's a good idea to put together an online portfolio to show potential employers the work you've done in the past—even if it was not professionally commissioned, it's an opportunity to show them your talent and range.

Writing Your Cover Letter

As well as your résumé, most employers will ask that you submit a cover letter. This is a one-page letter in which you express your motivation, why you are interested in the organization or position, and what skills you possess that make you the right fit.

Here are some tips for writing an effective cover letter:

- As always, proofread your text carefully before submitting it.
- Be sure you have a letter that is focused on a specific job. Do not make it too general or one-size-fits-all.
- Summarize why you are right for the position.
- Keep your letter to one page.
- Introduce yourself in a way that makes the reader want to know more about you and encourage them to review your résumé.
- Be specific about the job you are applying for. Mention the title and be sure it is correct.
- Try to find the name of the person who will receive your letter rather than keeping it nonspecific ("To whom it may concern").
- Be sure you include your contact details.
- End with a "call to action"—a request for an interview, for example.

Interviewing Skills

With your sparkling résumé and LinkedIn profile, you are bound to be called for an interview. This is an important stage to reach: you will have already gone through several filters—a potential employer has gotten a quick scan of your experience (remember, on average a résumé is viewed for only six seconds!) and has reviewed your LinkedIn profile and has made the decision to learn more about you in person.

There's no way to know ahead of time exactly what to expect in an interview, but there are many ways to prepare yourself. You can start by learning more about the person who will be interviewing you. In the same way recruiters and employers can learn about you online, you can do the same. You can see if you have any education or work experience in common, or any contacts you both know.

Preparing yourself for the types of questions you will be asked to ensure you offer a thoughtful and meaningful response is vital to interview success. Consider your answers carefully, and be prepared to support them with examples and anecdotes.

- Why did you decide to enter this field? What drives your passion for gaming?
- What is your educational background? What credentials did you earn?
- What did you like best about the education experience? What did you like least?
- Where and how were you trained?
- What is your management style? What management style do you prefer your supervisor to have?
- How many employees report to you? What levels are the employees who are your direct reports?
- Are you a team player? Describe your usual role in a team-centered work environment. Do you easily assume a leadership role?

Dressing Appropriately

How you dress for a job interview is very important to the impression you want to make. Remember that the interview, no matter what the actual environment in which you'd be working, is your chance to present your most professional self. Although you will not likely ever wear a suit to work, for the interview it's the most professional choice.

A job interview can be stressful, but with the right preparation you will interview with confidence. *iStock*

BEWARE WHAT YOU SHARE ON SOCIAL MEDIA

Most of us engage in social media. Sites such as Facebook, Twitter, and Instagram provide us a platform for sharing photos and memories, opinions and life events, and reveal everything from our political stance to our sense of humor. It's a great way to connect with people around the world, but once you post something, it's accessible to anyone—including potential employers—unless you take mindful precaution.

Your posts may be public, which means you may be making the wrong impression without realizing it. More and more, people are using search engines like Google to get a sense of potential employers, colleagues, or employees, and the impression you make online can have a strong impact on how you are perceived. According to the website Career Builder, 60 percent of employers search for information on candidates on social media sites.[5]

The website Glassdoor offers the following tips for how to avoid your social media activity from sabotaging your career success:

1. Check your privacy settings. Ensure that your photos and posts are only accessible to the friends or contacts you want to see them. You want to come across as professional and reliable.
2. Rather than avoid social media while searching for a job, use it to your advantage. Give future employees a sense of your professional interest by liking pages or joining groups of professional organizations related to your career goals.
3. Grammar counts. Be attentive to the quality of writing of all your posts and comments.
4. Be consistent. With each social media outlet, there is a different focus and tone of what you are communicating. LinkedIn is very professional while Facebook is far more social and relaxed. It's okay to take a different tone on various social media sites, but be sure you aren't blatantly contradicting yourself.
5. Choose your username carefully. Remember, social media may be the first impression anyone has of you in the professional realm.[6]

Hygiene is very important at any interview, but when you are seeking a job preparing and serving food it's paramount that you focus on being clean and well groomed. If you have long hair, it's a good idea to tie it back. If you wear makeup, keep it light and subtle.

Although you may be applying for a job in a casual, laid-back environment, until the job is yours it's important to come across as a professional, including dressing the part when you interview. A suit is no longer an absolute requirement, but avoid looking too casual, as that will give the impression you are not that interested.

What Employers Expect

Hiring managers and human resource professionals will also have certain expectations of you at an interview. The main thing is preparation: it cannot be overstated that you should arrive to an interview appropriately dressed, on time, unhurried, and ready to answer—and ask—questions.

For any job interview, the main things employers will look for are that you:

- Have a thorough understanding of the organization and the job for which you are applying.
- Be prepared to answer questions about yourself and your relevant experience.
- Be poised and likable, but still professional. They will be looking for a sense of what it would be like to work with you on a daily basis and how your presence would fit in the culture of the business.
- Stay engaged. Listen carefully to what is being asked and offer thoughtful but concise answers. Don't blurt out answers you've memorized, but really focus on what is being asked.

Be prepared to ask your own questions. It shows how much you understand the flow of an organization or workplace and how you will contribute to it. Some questions you can ask:

- What created the need to fill this position? Is it a new position or has someone left the company?
- Where does this position fit in the overall hierarchy of the organization?
- What are the key skills required to succeed in this job?
- What challenges might I expect to face within the first six months on the job?
- How does this position relate to the achievement of the company's (or department's, or boss's) goals?
- How would you describe the company culture?

MAKING GAMES FOR YOUR AUDIENCE, NOT YOURSELF

Dan Nanni. *Courtesy of Dan Nanni*

Dan Nanni, lead designer, manages the game and level design department at his studio. Part of his job is designing features for whatever video game he's working on; the other part is managing the other designers that report to him.

What is a typical day in your job?

I read, write, and talk. A lot. My role changes as the product I'm working on moves forward. They can often take two or more years from start to completion, and every major milestone my role shifts to adapt to the development process. Early on it's writing down designs in the form of pitches and talking to executives and directors about an idea I think will make our product special. Then, when we've narrowed down the direction of the game, it turns into technical design documents that provider coders and artists with a blueprint of whatever game features I'm designing. After that, when the features get implemented, it's going into the game engine and working on the features themselves, balancing them out for fun and fairness, and also figuring what additional work needs to be done to make sure it's something the player enjoys. In the end, it's fixing bugs, playing the game—a lot—to find out what is or isn't working right, and ultimately making sure that the game launches at the highest quality possible, while, hopefully, making something special in the process. During this time, I'm also managing my designers and collaborating with the other department leads and studio directors. I'm mentoring the designers that work with me, setting the schedule and providing them with direction, while also ensuring that the needs of other departments—such as art and code—along with the desires of our studio directors and executives, are well represented in whatever game designs we're working on.

What's the best part of your job?

Seeing a game reach the finish line. It's not an easy process, and there's a lot of stress along the way, but it's a great feeling to see something you've worked on very hard enjoyed by your fans. It doesn't always succeed, and that's not always due to lack of trying. On the contrary, you can often work above and beyond what should be expected of you, but still release something that didn't succeed. But there's still a ton of pride packed into that game, regardless of the outcome. It may not be something visible to the player, or anyone else besides you for that matter, but with every project that ships, a piece of you goes with it. And it's something you're proud of, regardless of the outcome.

What's the worst part of your job?

The unpredictability. I've been at this for about twenty years now. I've completed around ten major moves during my career, which span across ten cities, five states and three countries. Some companies have been fantastic, others have been soul crushing. It can be very hit or miss in this industry since you don't really know who's running the show. Not everyone who starts a game company knows what they're doing from a business perspective. They may have worked on several successful games, but the human side of running a business is lacking. Or the business side of running a business is lacking for that matter. Finding companies that understand how to balance making great games, with running a successful business, and also treating their employees fairly and professionally—that's not an easy thing to find in this industry.

What's the most surprising thing about your job?

How much you're constantly learning. It's not a surprise to me any longer, but it's not something I went into this industry expecting. You aren't just constantly learning about new technologies and practices, but you're also learning a lot about many different topics—geography, history, science, math, astronomy, writing, architecture, psychology, and the list goes on and on. From project to project the subject of your game often changes, and with it you'll be learning something new to ensure your players experience that in a meaningful way. But then there's the fact your players need to learn how to play the game, and so you'll start learning how humans interact with machines to trigger excitement, tap into intuition, and pass on vital information. It's very fun, but also very demanding. If you're not someone who's willing to adapt, this industry can seem very chaotic.

What's next, where do you see yourself going from here?

I like working with the human side of development, and would like to work more on managing employees. I love working on games, and as I get older I find that I like empowering people to make the games more than coming up with the specifics. I've had my fair share of coming up with exciting game systems and successful products, and I find that it's more rewarding sharing that experience with others and seeing them succeed. Maybe my perspective has shifted since I've become a father, but I like seeing the growth in individuals more than I like seeing the growth in a specific project these days. I also recognize that there's a whole new generation of game developers in this industry now that are more representative of the audience buying our games. What I like and prefer as a gamer isn't necessarily what everyone else is playing at home. As a designer it's really important that I'm making games for my audience, not for myself, and I rely on those I work with to help me find successful solutions to our design problems.

Did your education prepare you for the job?

Yes and no. I graduated with an art degree, and while my first job in the game industry was as an artist, I no longer practice it and haven't for many years. I couldn't get hired for my artistic skills anymore, but I can still have conversations with artists. Since design is heavily based on communication, being able to talk with artists in specifics helps with critiques and giving direction. So while my degree isn't something I use on a daily basis for actual work, it has given me a foundation that I was able to build my career on.

Is the job what you expected?

Mostly, yes. I was warned that it would be unpredictable and stressful. As someone who didn't like—many, many years ago—writing and history, I don't think there was any way you could have convinced me that those would be two important subjects I'd be using more often than not. And that I'd even grow to love them. I would have also loved it if someone told me that public speaking would have been something I should have planned for in the future. Game developers aren't the most graceful people in front of cameras, and I would have really appreciated a heads-up on this. I was also warned that I'd eventually stop playing games in my free time as I'd start to burn out on them. This isn't necessarily true. I still like playing games, but it comes in bursts. I might have a drought for a few months, especially as work really starts picking up. But when I find a game that I love, I can still play for hours at a time. I just can't do it as often as I used to. And I'm also very impatient with games—if something turns me off within the first thirty minutes, then I can't keep playing it without it feeling like work. My tolerance for what turns me off is very low as well. All in all though, it's been an exciting industry to work in. Not always a good, exciting ride, but it has been a crazy ride, and I've had several wonderful experiences along the way. If given the option, I'd do it again, but I don't think I would have approached it from an art background if I did it again. Knowing what I know now, and what I like doing, I think I would have rather approached it from a focus in history, psychology, or creative/business management perspective.

Summary

Congratulations on working through the book! You should now have a strong idea of your career goals within the computer gaming industry and how to realize them. In this chapter, we covered how to present yourself as the right candidate to a potential employer—and these strategies are also relevant if you are applying to a college or another form of training.

Here are some tips to sum it up:

- Your résumé should be concise and focused on only relevant aspects of your work experience or education. Although you can include some personal hobbies or details, they should be related to the job and your qualifications for it.
- Take your time with all your professional documents—your résumé, your cover letter, your LinkedIn profiles—and be sure to proofread very

carefully to avoid embarrassing and sloppy mistakes.

- Prepare yourself for an interview by anticipating the types of questions you will be asked and coming up with professional and meaningful responses.

- Equally, prepare some questions for your potential employer to ask at the interview. This will show you have a good understanding and interest in the organization and what role you would have in it.

- Always follow up after an interview with a letter or an e-mail. An e-mail is the fastest way to express your gratitude for the interviewer's time and restate your interest in the position.

- Dress appropriately for an interview and pay extra attention to tidiness and hygiene.

- Be wary of what you share on social media sites while job searching. Most employers research candidates online and what you have shared will influence their idea of who you are and what it would be like to work with you.

The computer gaming industry is an exciting one with many different types of jobs and work environments. This book has described the various jobs and provided examples of real working professionals and their impressions of what they do and how they prepare—through education or training—to do it. We hope this will further inspire you to identify your goal and know how to achieve it.

You've chosen a field that is expected to grow in the coming years and one that will offer a creative, diverse, competitive, and exciting career path. People will always enjoy interacting technology and, through it, with each other, and playing challenging and fun games that entertain and make them think. We wish you great success in your future.

Notes

Introduction

1. CNBC. "Video Game Industry Is Booming with Continued Revenue," July 18, 2018, https://www.cnbc.com/2018/07/18/video-game-industry-is-booming -with-continued-revenue.html.

2. Daniel Greenspan, "Best Computer Jobs for the Future," IT Career Finder, March 19, 2019, https://www.itcareerfinder.com/brain-food/blog/entry/best-computer -jobs-for-the-future.html.

3. John Ballard, "5 Trends Explain the Growth of the Video-Game Industry," *Motley Fool*, November 9, 2008, https://www.fool.com/investing/2018/11/09/5 -trends-explain-the-growth-of-the-video-game-indu.aspx.

4. WePC, "2019 Video Game Industry Statistics, Trends & Data," April 2019, https://www.wepc.com/news/video-game-statistics/.

5. Ballard, "5 Trends Explain the Growth of the Video Game Industry."

6. Craig Casazza, "Best Cities to Be a Game Developer," *ValueGamers*, September 18, 2018, https://valuegamers.com/analysis/best-cities-to-be-a-game-developer.

Chapter 1

1. History.com, "Video Game History," August 21, 2018, https://www.history .com/topics/inventions/history-of-video-games.

2. The Douglas Archives, "Professor Sandy Douglas," http://www.douglashistory .co.uk/history/sandy_douglas.htm.

3. Computer History Museum, "Steve Russell," https://www.computerhistory.org /events/bio/Steve,Russell.

4. Mary Bellis, "The History of the UNIVAC Computer," ThoughtCo, March 5, 2019, https://www.thoughtco.com/the-history-of-the-univac-computer-1992590.

5. The Strong National Museum of Play, "Video Game History Timeline," https:// www.museumofplay.org/about/icheg/video-game-history/timeline.

6. BugSplat Editors, "What Was the Great Video Game Crash of 1983?" BugSplat, April 22, 2018, https://www.bugsplat.com/articles/video-games/great-video -game-crash-1983.

7. Debate.org, "Is the Video Game Industry on the Verge of Another Market Crash?" https://www.debate.org/opinions/is-the-video-game-industry-on-the-verge-of -another-market-crash.

8. Writers Guild, "2019 Writers Guild Awards Winners and Nominees," February 17, 2018, https://awards.wga.org/awards/nominees-winners.

9. Bureau of Labor Statistics, "Multimedia Artists and Animators," https://www .bls.gov/ooh/arts-and-design/multimedia-artists-and-animators.htm.

10. Bureau of Labor Statistics, "Producers and Directors," https://www.bls.gov /ooh/entertainment-and-sports/producers-and-directors.htm.

11. Bureau of Labor Statistics, "Broadcast and Sound Engineering Technicians," https://www.bls.gov/ooh/media-and-communication/broadcast-and-sound-engineer ing-technicians.htm.

12. Bureau of Labor Statistics, "Software Developers," https://www.bls.gov/ooh /computer-and-information-technology/software-developers.htm.

13. Bureau of Labor Statistics, "Computer Programmers," https://www.bls.gov /ooh/computer-and-information-technology/computer-programmers.htm.

14. Guru 99, "Software Testing as a Career," https://www.guru99.com/software -testing-career-complete-guide.html.

15. Bureau of Labor Statistics, "Writers and Authors," https://www.bls.gov/ooh /media-and-communication/writers-and-authors.htm.

Chapter 2

1. Jason W. Bay, "How to Become a Video Game Tester (FAQ)," Game Industry Career Guide, https://www.gameindustrycareerguide.com/how-to-become -a-video-game-tester/.

2. Sheryl Burgstahler, Sara Lopez, and Scott Bellman, "Preparing for a Career: An Online Tutorial," DO-IT, https://www.washington.edu/doit/preparing-career -online-tutorial.

3. 80 Level, "10 Insider Tips for Artists Applying to Game Studios," March 9, 2018, https://80.lv/articles/10-insider-tips-for-artists-applying-to-game-studios/.

4. Jason W. Bay, "Is Your Online Art Portfolio Lacking? Get Advice, Inspiration from 7 Professionals," Game Industry Career Guide, https://www.gameindustrycareer guide.com/building-artist-portfolio-site/.

5. Study.com, "Be a Game Artist: Requirements and Information," https://study.com/articles/Be_a_Game_Artist_Requirements_and_Information.html.

6. Study.com, "Video Game Producer: Education Requirements and Career Info," https://study.com/articles/Video_Game_Producer_Education_Requirements_and_Career_Info.html.

7. Study.com, "Sound Designer: Education Requirements and Career Info," https://study.com/articles/Sound_Designer_Education_Requirements_and_Career_Info.html.

8. Study.com, "How to Become a Video Game Designer: Education and Career Roadmap," https://study.com/articles/How_to_Become_a_Video_Game_Designer_Education_and_Career_Roadmap.html.

9. Study.com, "Game Developer: Job Description, Duties and Requirements," https://study.com/articles/Game_Developer_Job_Description_Duties_and_Requirements.html.

10. Study.com, "How to Become a Videogame Tester," https://study.com/articles/Information_on_Becoming_a_Video_Game_Tester.html.

Chapter 3

1. Steven R. Antonoff, "College Personality Quiz," *U.S. News & World Report*, May 15, 2019, https://www.usnews.com/education/best-colleges/right-school/choices/articles/college-personality-quiz.

2. Princeton Review, "Top 50 Game Design: Ugrad," https://www.princetonreview.com/college-rankings?rankings=top-50-game-design-ugrad.

3. College Rank, "The 20 Best Animation Degree Programs," https://www.collegerank.net/best-animation-degree-programs/.

4. *U.S. News & World Report*, "Best Computer Science Schools," 2018, https://www.usnews.com/best-graduate-schools/top-science-schools/computer-science-rankings.

5. Caroline Koppelman, "The Best Undergraduate Creative Writing Programs," Koppelman Group, November 5, 2017, https://www.koppelmangroup.com/blog/2017/11/5/the-best-undergraduate-creative-writing-programs.

6. US Department of Education, Federal Student Aid, https://studentaid.ed.gov/sa/fafsa.

Chapter 4

1. Jason W. Bay, "Real Résumé Examples That Worked," Game Industry Career Guide, https://www.gameindustrycareerguide.com/videogame-resume-example/.

2. Sarah E. Needleman, "When a Passion for Videogames Helps Land That Job," *Wall Street Journal*, March 6, 2019, https://www.wsj.com/articles/when-a-passion-for-videogames-helps-land-that-job-11551888001.

3. The Balance Careers, "Student Resume Examples and Templates," https://www.thebalancecareers.com/student-resume-examples-and-templates-2063555.

4. Joshua Waldman, *Job Searching with Social Media for Dummies* (Hoboken, NJ: Wiley, 2013), 148–49.

5. Career Builder, "Number of Employers Using Social Media to Screen Candidates Has Increased 500 Percent over the Last Decade," April 28, 2016, http://www.careerbuilder.com/share/aboutus/pressreleasesdetail.aspx?ed=12%2F31%2F2016&id=pr945&sd=4%2F28%2F2016.

6. Alice E. M. Underwood, "9 Things to Avoid on Social Media While Looking for a New Job," Glassdoor, January 3, 2018, https://www.glassdoor.com/blog/things-to-avoid-on-social-media-job-search/.

Glossary

2-D animation: A type of animation in which the images are "flat," meaning they have width and height but no depth.

3-D animation: A type of animation in which images appear in a three-dimensional space, with width, height, and depth.

animation: The art of creating electronic images with a computer in order to create moving images.

artist/animator: The person who creates the environments and other animated, interactive images for computer games.

bachelor's degree: A four-year degree awarded by a college or university.

bugs: In the context of computing, a bug is a glitch or error in the code or in how a computer program acts or responds. Testers are responsible for discovering bugs in programs including games.

burnout: Feeling of physical and emotional exhaustion caused by overworking.

business model: A map for the successful creation and operation of a business, including sources of revenue, target customer base, products, and details of financing.

campus: The location of a school, college, or university.

career assessment test: A test that asks questions particularly geared to identify skills and interests to help inform the test taker of what type of career would suit them.

colleagues: The people with which you work alongside.

community college: A two-year college that awards associate's degrees.

computer science: The study of the principles and use of computers, such as programming.

cover letter: A document that usually accompanies a résumé and allows a candidate applying to a job or a school or internship an opportunity to describe their motivation and qualifications.

deadlines: Targets set relating to when a particular task needs to be completed. In computer gaming, deadlines are often extremely tight and change frequently.

educational background: The schools attended and degrees a person has earned.

entrepreneur: A person who creates, launches, and manages his or her own business.

financial aid: Various means of receiving financial support for the purposes of attending school. This can be a grant or scholarship, for example.

freelancer: A person who owns his or her own business that provides services for a variety of clients.

game console: The hardware on which computer games are played, such as the Atari and Sega System.

game developer: The more technical side of video game creation, a game developer is a computer programmer or engineer who can take a game designer's vision and perform the coding and programming to bring it to life.

game designer: A person responsible for conceptualizing a new game, including the characters, story line, setting for the game, and rules of play.

game tester: A video game tester is hired by video game companies to ensure games work as they should before being released, testing each possible action or movement in a game to ensure there are no bugs.

gap year: A year between high school and higher education or employment during which a person can explore his or her passions and interests, often while traveling.

general educational development (GED): A certificate earned by someone who has not graduated from high school that is the equivalent to a high school diploma.

industry: The people and activities involved in one type of business, such as the business of creating computer games.

internship: A work experience opportunity that lasts for a set period of time and can be paid or unpaid.

interpersonal skills: The ability to communicate and interact with other people in an effective manner.

interviewing: A part of the job-seeking process in which a candidate meets with a potential employer, usually face to face, in order to discuss his or her work experience and education and seek information about the position.

job market: A market in which employers search for employees and employees search for jobs.

major: The subject or course of study in which one chooses to earn a degree.

master's degree: A degree that is sought by those who have already earned a bachelor's degree in order to further their education.

multimedia art: Artwork created with new media technologies and computers.

networking: The processes of building, strengthening, and maintaining professional relationships as a way to further your career goals.

on-the-job training: A type of training in which a person is learning while actually doing the job in a real-world environment.

OXO: The very first computer game ever developed, which appeared in 1952 and was based on the game tic-tac-toe.

portfolio: A collection of work, be it writing, animation, or any other form of creative output, that represents the talents and abilities of the artist.

producer: A person who oversees a team of engineers and artists and all others involved in the creation of a game, and ensuring the project is on track.

résumé: A document, usually one page, that outlines a person's professional experience and education that is designed to give potential employers a sense of a candidate's qualifications.

sculpting (or digital sculpting): The act of creating 3-D objects with a computer using a material similar to clay.

scriptwriting: The creation of commands that are followed by computer programs are scripting engines.

social media: Websites and applications that enable users to create and share content online for networking and social sharing purposes. Examples include Facebook and Instagram.

sound designer: The person responsible for all the sounds in a game, including hiring and recording voice actors for characters and creating the background and actions sound effects in a game.

Spacewar!: The first game that was possible to play on more than one computer installation.

storyboarding: A sequence of drawings, accompanied by directions and dialogue, that represent shots planned for a computer game or any other type of video or film production.

technology hubs: Metropolitan areas with a dense presence of technical companies.

tuition: The money you have to pay for education, be it a university degree or a certification.

UNIVAC I (UNIVersal Automatic Computer I): The first commercially available computer.

university labs: Spaces in universities where experiments and research take place.

vocational school: A school in which students learn how to do a job that requires special skills.

work culture: A concept that defines the beliefs, philosophies, thought processes, and attitudes of employees in a particular organization.

writer: The person who creates the dialogue, develops the plot, and works on character creation.

ZBrush: A digital sculpting tool that combines 3-D/2.5-D modeling, texturing, and painting.

Resources

The following websites, magazines, and organizations can help you further investigate and educate yourself on topics related to computer gaming, all of which will help you as you take the next steps in your career, now and throughout.

Professional Organizations

Digital Games Research Association
http://www.digra.org/
An international association for academics and professionals who research digital games.

Entertainment Software Association
http://www.theesa.com/
An association focused on the business and on companies that create computer and video games for video game consoles, handheld devices, personal computers, and the internet.

International Game Developers Association
https://www.igda.org/
A nonprofit professional association and a global network of professionals from all fields of game development, from programmers and producers to designers and writers—and everyone else who participates in any way in the game development process.

SIGCHI
https://sigchi.org/
An international organization of academics, professionals, and students who are focused on human-technology and human-computer interaction. The organization also offers workshops, tutorials, and various forms of outreach.

Peace Corps
https://www.peacecorps.gov/

If you are interested in taking a gap year before taking the next step in your career or education, consider joining the Peace Corps. Volunteers have the experience of working on projects that relate to health, agriculture, education, and youth and development, just to name a few, and the experience can help you find your passion and understand what the next step in your life should be.

Magazines

Game Informer
https://www.gameinformer.com

The world's largest gaming magazine; a source for the latest in video game news, reviews, previews, podcasts, and features.

Old School Gamer
https://oldschoolgamermagazine.com

A magazine dedicated to gamers who are enthusiasts about classic arcade games.

PC Gamer
https://www.pcgamer.com

The global authority on PC games has been reporting on the subject for more than twenty years. There are worldwide print editions and around-the-clock news, features, e-sports coverage, hardware testing, and game reviews available on the website.

Pocket Tactics
https://www.pockettactics.com

A magazine based in England about strategy gaming (and other games for grownups) for Android and iOS devices, updated every day.

Retro Gaming
http://retrogamingmagazine.com

Covers all retro gaming content no matter the platform, from Atari 2600 to Windows, from Super Nintendo to Commodore Amiga.

Retro Gamer

https://www.retrogamer.net

Sure to inspire enthusiasm for the games it reports on, this is a magazine fully dedicated to the golden days of classic gaming, keeping you up to date with informative and in-depth stories and access to legendary developers.

Bibliography

Antonoff, Steven R. "College Personality Quiz." *U.S. News & World Report*, May 15, 2019. https://www.usnews.com/education/best-colleges/right-school /choices/articles/college-personality-quiz.

80 Level. "10 Insider Tips for Artists Applying to Game Studios," March 9, 2018. https://80.lv/articles/10-insider-tips-for-artists-applying-to-game-studios/.

The Balance Careers. "Student Resume Examples and Templates." https://www .thebalancecareers.com/student-resume-examples-and-templates-2063555.

Ballard, John. "5 Trends Explain the Growth of the Video-Game Industry," *Motley Fool*, November 9, 2008. https://www.fool.com/investing/2018/11/09/5 -trends-explain-the-growth-of-the-video-game-indu.aspx.

Bay, Jason W. "How to Become a Video Game Tester (FAQ)." Game Industry Career Guide. https://www.gameindustrycareerguide.com/how-to-become -a-videogame-tester/.

———. "Is Your Online Art Portfolio Lacking? Get Advice, Inspiration from 7 Professionals." Game Industry Career Guide. https://www.gameindustry careerguide.com/building-artist-portfolio-site/.

———. "Real Resume Examples That Worked." Game Industry Career Guide." https://www.gameindustrycareerguide.com/video-game-resume-example/.

Bellis, Mary. "The History of the UNIVAC Computer." ThoughtCo, March 5, 2019. https://www.thoughtco.com/the-history-of-the-univac-computer -1992590.

BugSplat Editors. "What Was the Great Video Game Crash of 1983?" BugSplat, April 22, 2018. https://www.bugsplat.com/articles/video-games /great-video-game-crash-1983.

Bureau of Labor Statistics. "Broadcast and Sound Engineering Technicians." https://www.bls.gov/ooh/media-and-communication/broadcast-and -sound-engineering-technicians.htm.

———. "Computer Programmers." https://www.bls.gov/ooh/computer-and -information-technology/computer-programmers.htm.

———. "Multimedia Artists and Animators." https://www.bls.gov/ooh/arts-and-design/multimedia-artists-and-animators.htm.

———. "Software Developers." https://www.bls.gov/ooh/computer-and-information-technology/software-developers.htm.

———. "Producers and Directors." https://www.bls.gov/ooh/entertainment-and-sports/producers-and-directors.htm.

———. "Writers and Authors." https://www.bls.gov/ooh/media-and-communication/writers-and-authors.htm.

Burgstahler, Sheryl, Sara Lopez, and Scott Bellman. "Preparing for a Career: An Online Tutorial." DO-IT. https://www.washington.edu/doit/preparing-career-online-tutorial.

Career Builder. "Number of Employers Using Social Media to Screen Candidates Has Increased 500 Percent over the Last Decade," April 28, 2016. https://www.careerbuilder.com/share/aboutus/pressreleasesdetail.aspx?ed=12%2F31%2F2016&id=pr945&sd=4%2F28%2F2016.

Casazza, Craig. "Best Cities to Be a Game Developer." *ValueGamers*, September 18, 2018. https://valuegamers.com/analysis/best-cities-to-be-a-game-developer.

CNBC. "Video Game Industry Is Booming with Continued Revenue," July 18, 2018. https://www.cnbc.com/2018/07/18/video-game-industry-is-booming-with-continued-revenue.html.

Cohen, D. S. "History of the Atari 2600 VCS." Lifewire, November 25, 2018. https://www.lifewire.com/atari-2600-console-729665.

College Rank. "The 20 Best Animation Degree Programs." https://www.collegerank.net/best-animation-degree-programs/.

Computer History Museum. "Steve Russell." https://www.computerhistory.org/events/bio/Steve,Russell.

Debate.org. "Is the Video Game Industry on the Verge of Another Market Crash?" https://www.debate.org/opinions/is-the-video-game-industry-on-the-verge-of-another-market-crash.

The Douglas Archives. "Professor Sandy Douglas." http://www.douglashistory.co.uk/history/sandy_douglas.htm.

Ell, Kellie. "Video Game Industry Is Booming with Continued Revenue." CNBC, July 8, 2018. https://www.cnbc.com/2018/07/18/video-game-industry-is-booming-with-continued-revenue.html.

Greenspan, Daniel. "Best Computer Jobs for the Future." IT Career Finder, March 19, 2019. https://www.itcareerfinder.com/brain-food/blog/entry /best-computer-jobs-for-the-future.html.

Guru99. "Software Testing as a Career." https://www.guru99.com/software -testing-career-complete-guide.html.

History.com. "Video Game History," August 21, 2018. https://www.history .com/topics/inventions/history-of-video-games.

Koppelman, Caroline. "The Best Undergraduate Creative Writing Programs." Koppelman Group, November 5, 2017. https://www.koppelmangroup .com/blog/2017/11/5/the-best-undergraduate-creative-writing-programs.

Needleman, Sarah E. "When a Passion for Videogames Helps Land That Job." *Wall Street Journal*, March 6, 2019. https://www.wsj.com/articles /when-a-passion-for-videogames-helps-land-that-job-11551888001.

Princeton Review. "Top 50 Game Design: Ugrad." https://www.princeton review.com/college-rankings?rankings=top-50-game-design-ugrad.

The Strong National Museum of Play. "Video Game History Timeline." https:// www.museumofplay.org/about/icheg/video-game-history/timeline.

Study.com. "Be a Game Artist: Requirements and Information." https://study .com/articles/Be_a_Game_Artist_Requirements_and_Information.html.

———. "Game Developer: Job Description, Duties and Requirements." https://study.com/articles/Game_Developer_Job_Description_Duties _and_Requirements.html.

———. "How to Become a Video Game Designer: Education and Career Roadmap." https://study.com/articles/How_to_Become_a_Video_Game _Designer_Education_and_Career_Roadmap.html.

Study.com. "How to Become a Videogame Tester." https://study.com/articles /Information_on_Becoming_a_Video_Game_Tester.html.

Study.com. "Sound Designer: Education Requirements and Career Info." https://study.com/articles/Sound_Designer_Education_Requirements _and_Career_Info.html.

———. "Video Game Producer: Education Requirements and Career Info." https://study.com/articles/Video_Game_Producer_Education_Require ments_and_Career_Info.html.

Underwood, Alice E. M. "9 Things to Avoid on Social Media While Looking for a New Job." Glassdoor, January 3, 2018. https://www.glassdoor.com /blog/things-to-avoid-on-social-media-job-search/.

U.S. News & World Report. "Best Computer Science Schools," 2018. https://
    www.usnews.com/best-graduate-schools/top-science-schools/computer
    -science-rankings.

Waldman, Joshua. *Job Searching with Social Media for Dummies.* Hoboken, NJ:
    Wiley, 2013.

WePC. "2019 Video Game Industry Statistics, Trends & Data," April 2019.
    https://www.wepc.com/news/video-game-statistics/.

Writers Guild. "2019 Writers Guild Awards Winners and Nominees," February
    17, 2019. https://awards.wga.org/awards/nominees-winners.

About the Author

Tracy Brown Hamilton is a writer, editor, and journalist based in the Netherlands. She has written several books on topics ranging from careers to media, economics to pop culture. She lives with her husband and three children.